GLOBAL WARMING:
THE GREAT DECEPTION

GLOBAL WARMING:
THE GREAT DECEPTION

The Triumph of Dollars and Politics Over Science and Why You Should Care

---

GUY K. MITCHELL, JR.

Global Warming: The Great Deception

Published in Savannah, Georgia by Literary Management Group

ISBN 979-8-7908-2994-9

Book Cover and Text design by BluDesign, Nashville, TN

Printed in the United States of America

CONTENTS

DEDICATION

I WOULD LIKE TO DEDICATE this book to Dr. Milan S. Djordjevic, PhD (1913 – 2000). Born in Belgrade, Yugoslavia, Dr. Djordjevic received his B.S. in Mechanical Engineering from the University of Belgrade in 1936 and a PhD in Mechanical Engineering from the University of Munich in 1939. He fought for the Allies in WWII and spent two years as a German prisoner of war at Stalag 16. He was Professor of Mechanical Engineering at Belgrade University, Duke University, and the University of Alabama-Tuscaloosa, from which he retired in 1981.

Dr. Djordjevic was a strong proponent of the scientific method of inquiry. His insistence on disciplined, logical thought and methodological rigor in analyzing a problem made an impression on me that has lasted a lifetime. He had the mind of a scientist and was an expert in the practical application of that science in the field. In the classroom, he could explain the thermodynamic processes involved in the operation of the Otto Cycle – the idealized thermodynamic model of a spark ignition engine. Yet, as an expert in the operation of internal combustion engines and a consultant to industry, he had the practical knowledge to go into the field and troubleshoot a problem with a large, stationary internal combustion engine.

Dr. Djordjevic taught many of my mechanical engineering courses and I considered him to be my mentor. I am grateful for the time he spent tutoring me on many subjects, but mostly teaching me the value of critical thinking. I would like to think that he would approve of my approach in writing this book.

ACKNOWLEDGMENTS

I WANT TO EXPRESS my appreciation to Dr. Roy Clark, who spent innumerable hours tutoring me in the basics of spectroscopy, quantum mechanics and atmospheric physics as relates to the subject of anthropogenic global warming. In addition, Roy reviewed all of the technical aspects of this book and provided valuable suggestions to help "get the science right."

Dr. Clark is a retired engineer with over 30 years of experience in new product and process development with an emphasis on optical and spectroscopic measurements in adverse environments.

He has successfully integrated complex laser diagnostics into large scale hypersonic and high-energy laser test facilities. He has also developed LED and fiber optic illumination systems and sensors for a wide range of applications. His spectroscopic experience extends from 200 nm to 200 cm^{-1} including work with circular and linearly polarized light. He has eight United States patents and thirteen technical publications. He received his M.A. in chemistry from Oxford University and his Ph.D. in chemical physics from Sussex University, U.K., in 1976.

He started his own research on climate change in 2007. His particular interest is time dependent or dynamic surface energy transfer, and the calculation of surface temperatures from first principles. He has published several technical articles on climate change including, "A Null Hypothesis for CO_2" and a book entitled, *The Dynamic Greenhouse Effect and the Climate Average Paradox: Ventura Photonics Monograph VPM 001.*

I would like to express my appreciation to my good friends Stewart Dansby, Pat Keeley, and James Morphy for the time they spent reviewing the manuscript for this book. Their input was invaluable in my efforts to improve the clarity of my writing. I want to add my special thanks to Delambert (Stowe) Rose for his review of the original manuscript and contributions to articles that were published in the *North State Journal* based on the content of this book. I want to give special recognition to Christina Sullivan, who served as my editor throughout the writing of this book. Chris spent countless hours checking references, formatting, and preparing this book for publication. It would not have happened without her considerable efforts. Finally, I want to thank Bruce Barbour, who provided the final editing and managed the publication of this book.

FOREWORD

I BEGAN MY OWN RESEARCH into anthropogenic global warming (AGW) in 2007. At that time, I had over 30 years of experience in optics and spectroscopy, including new product and process development for adverse environments such as oil well logging, high energy laser systems and hypersonic combustion. I started with a simple question: how did an increase in the atmospheric concentration of CO_2 change the surface temperature of the Earth? I did not find a satisfactory answer in the Intergovernmental Panel on Climate Change (IPCC) discussions of radiative forcings, feedbacks and climate sensitivities.

Therefore, I started my own analysis of time dependent, or dynamic surface energy transfer, and the calculation of surface temperatures from first principles. I realized early on that the problem with the climate models was the underlying assumptions used to simplify the energy transfer calculations. Those simplifications created AGW as a mathematical artifact in the model calculations. Physical reality has been abandoned in favor of mathematical simplicity. I have published several technical articles including two on "Dynamic Thermal Reservoirs" and a book entitled, *"The Dynamic Greenhouse Effect and the Climate Average Paradox,"* Ventura Photonics Monograph VPM 001. None of these publications have had any effect on the simplified 'climate science' used by the IPCC and the climate modelers. The proponents of the AGW hypothesis refuse to consider any scientific analysis that challenges the "consensus."

I have particularly enjoyed discussing AGW with Guy Mitchell. His research into the effect of external influences on AGW has been very helpful in explaining the persistence of the AGW 'science'. I am in total agreement with his analysis of the flaws in the science of the AGW hypothesis and the premise that the science is settled. It has been corrupted to advance a socio-economic agenda. Mankind is not the primary source of climate change. Policies such as phasing out fossil fuels, taxing carbon emissions, changing to solar and wind energy, using more electric vehicles and other such futile schemes will not reduce atmospheric or ocean temperatures.

I highly recommend you read and study Guy's work, so you are equipped with the scientific information you need to challenge the arguments in favor of AGW. When will you be forced into energy poverty where you must choose between food or winter heat for your house? The 2021 winter power failures in Texas are just a foretaste of what lies ahead.

—Roy Clark, Ph.D.

INTRODUCTION

THERE ARE FOUR main topics that impact the subject of anthropogenic (man-made) global warming (AGW): science, research funding, politics, and global economics. Each plays an important role in how the subject has evolved over time within the environmentalist/climate science community to a cause célèbre. Today, politicians, actors, businessmen, athletes, and others in the public arena are advocates for the man-made global warming hypothesis and the elimination of fossil fuels. In most cases, I suspect that they are not familiar with the scientific facts behind the claims of man-made global warming.

The primary objective of this book is to refute the pseudo-science that underpins the AGW hypothesis, using the concept known as "first principles of science" to do so. The term first principles of science refers to the concept that theoretical investigations in scientific fields begin at the level of established science, *ab initio.* They do not make assumptions using empirical models and parameter fitting such as occurs in climate model predictions. As a result, I have gone to some depths in certain chapters dealing with the science to explain the first principles of the science that governs the processes under investigation. While the targeted reader of this work is the person who does not have the technical background to evaluate the claims of the proponents of the AGW hypothesis, I understand that some of the technical subject matter may be difficult to comprehend. If this is the case, I apologize. However, I believed it necessary to provide sufficient technical information and scientific evidence to provide the reader with the facts to support the rejection of the anthropogenic global warming hypothesis. The challenge comes in trying to find the right balance between the two. Hopefully, I have struck that balance.

A secondary, but equally important objective of this book, is to expose the true motivation behind the promotion of the AGW hypothesis by the United Nations Intergovernmental Panel on Climate Change ("UNIPCC"), certain world politicians and international investment banking firms.

I have received no financial support from any source while writing this book. I have no affiliations with the climate science community, industry, academia, or any organization that might promote or oppose the man-made global warming hypothesis.

Guy K. Mitchell, Jr.
June 2021

CHAPTER ONE

ANTHROPOGENIC GLOBAL WARMING

THE SCIENCE INVOLVED in the promotion of the AGW hypothesis has departed significantly from the *Scientific Method of Inquiry*, which has guided scientific research for over 400 years. The proponents of the AGW hypothesis within the climate science community appear to have accepted it as a *fait accompli*. The goal is not to confirm or falsify the hypothesis; it is to develop a computer model that can accurately predict the extent of global warming in the future based on the predicted concentration of CO_2 in the atmosphere. The concept of an average equilibrium temperature of the Earth's surface is an abstraction; it is a figment of the climate scientists' imagination, conjured up to support a fraudulent hypothesis. A calculation of equal value would be to determine the average zip-code in the U.S. to locate the average American city. Even so, the NOAA land surface temperature database depicts only a 0.014°F/yr average temperature increase over the last 140 years, hardly a heat wave and well within the measurement margin of error. The temperature anomaly (monthly change compared to the long-term average) in the lower troposphere is measured by NOAA satellites orbiting the Earth using microwave sounding technology. During the last 40 years, the temperature anomaly increase has been *de minimus*. Since 2000, the temperature of the world's oceans has been measured using the ARGO Float program; the data shows that the temperature change in the world's oceans is virtually non-existent.

Water vapor (H_2O) is the dominant greenhouse gas in moderating the Earth's climate; carbon dioxide (CO_2) plays virtually no role. A CO_2 molecule does not immediately reradiate an absorbed photon; it transfers that heat energy through molecular collision to a nearby molecule in the atmosphere. Only a small portion of the energy (less than 0.2% of the Sun's irradiance on a clear summer's day at equatorial latitudes) that is absorbed is reradiated downwards towards the Earth's surface after the molecular collision occurs. The world's oceans cannot absorb long-wave infrared (LWIR) photons below a depth of 100 microns-about the diameter of a human hair. Therefore, emissions from CO_2 molecules from the atmosphere can have no practical effect on the temperature of the world's oceans, which comprise 71% of the Earth's surface.

The world's oceans drive the Earth's climate. The amount of solar irradiance the ocean receives is responsible for its heat content. That value is a function of the geometry of the Earth's orbit, the sunspot cycle, and local atmospheric conditions. Cloud cover, relative humidity and surface wind speeds affect the ocean's ability to absorb sunlight and retain the absorbed heat. Cyclical variations of 60 – 80 years in the Atlantic Multi-decadal Oscillation causes cyclical melting of polar ice. Those cycles have nothing to do with man's activities on the Earth. The claims that LWIR photons heat either the Earth's surface or the world's oceans are refuted by well-established science such as thermodynamics, atmospheric physics, quantum mechanics and spectroscopy. Those claims are the product of flawed science funded by various organizations to promote the AGW hypothesis.

The stated goal of the United Nations Intergovernmental Panel on Climate Change (UN IPCC) is to address worldwide socio-economic inequality within the context of climate change. To accomplish that end, the UN IPCC has promoted a flawed scientific hypothesis regarding the role of CO_2 in the Earth's atmosphere that predicts global warming with dire consequences as the concentration of CO_2 increases. The computer models used by the UN IPCC consistently predict global warming that substantially exceeds that which occurs in reality. Every effort to demonstrate the efficacy of the predictive climate models by using "back testing" (predicting actual historical temperatures using actual historical data) fails. An accurate database containing the necessary climate parameters to populate the computer models with reliable data does not exist. The thermodynamic processes that affect the Earth's climate are too complicated to model with any degree of accuracy.

U.N. climate treaties have consistently failed to reduce CO_2 emissions into the atmosphere. During the term of the failed Kyoto Protocol, world emissions of CO_2 increased by 32% from 1990 – 2010.[1] The Paris Climate Accord does not stipulate mandatory reductions in CO_2 emissions for signatories nor does it have an enforcement mechanism. The results of the Paris Climate Accord will be no different than the failed Kyoto Protocol. The real purpose of climate treaties is to perpetuate the trading of carbon credits to enrich global investment firms. U.N. climate scientists' predictions of the amount of future global warming using flawed computer models, employing the concept of the Earth's average temperature, consistently prove to be grossly inaccurate. Even the flawed temperature database developed by climate scientists for the Earth's land mass, oceans and atmosphere demonstrates that there has been no significant global warming.

The amount of research funding provided by governments and non-profits around the world to advance the global warming hypothesis is staggering. It is estimated that worldwide, climate research funding has exceeded a trillion dollars.[2] The U.S. Government Accountability Office reported that U.S. government spending on climate change was about $13 billion in 2016; only around $5 billion was spent on cancer research.[3] On June 1, 2017, when President Trump announced that the U.S. was withdrawing from the Paris Climate Accord, he stated that the U.S. would decommit $2 billion of the $3 billion that President Obama had pledged to the Green Climate Fund. The Green Climate Fund was established by the U.N. in 2010 to "help developing countries reduce their greenhouse gas emissions and enhance their ability to respond to climate change." [4] The Green Climate Fund is nothing more than a vehicle to transfer wealth and enrich those who participate in the process.

The pressure on research scientists to affirm the hypothesis is relentless. Groupthink, in response to political pressure, has evolved among the universities and academic societies in the West such as MIT, NASA, the British Royal Society, the German Helmholtz Association and the U.S. National Academy of Sciences. It would be career suicide for a young climate scientist or academic to oppose the hypothesis. Not only would he or she see funding dry up, but they would be ostracized within the climate science community for breaking ranks.

Many politicians in Europe and the U.S. have joined the AGW chorus with full-throated support; it has assumed the element of noblesse oblige. "Even if the theory of global warming is wrong, we will be doing the right thing -- in terms of economic policy and environmental policy." ~ Tim Wirth, former U.S. Senator from Colorado. "No matter if the science is all phony, there are collateral environmental benefits.... climate change [provides] the greatest chance to bring about justice and equality in the world." ~ Christine Stewart, former Minister of the Environment of Canada. "There is a clear moral imperative for developed economies like the U.K. to help those around the world who stand to lose most from the consequences of man-made climate change." ~ British Prime Minister Theresa May in 2018 [5] The implication is that those who do not support the AGW hypothesis are immoral. In fact, the cause of man-made global warming has assumed a religious fervor among many of those who support it.

If all of this is true (and it is), why do the U.N. and certain scientists and politicians, particularly in the U.S. and Western Europe, continue to promote this fraudulent hypothesis? Why do they continue to posture global warming as an "existential threat"? The purpose of this book is to prove the foregoing statements and answer these questions.

CHAPTER TWO

97% OF THE WORLD'S SCIENTISTS

IN JUNE 2017, I attended a dinner party hosted by friends at their home in the North Carolina mountains. During the course of the conversation surrounding dinner, I asked a friend, who was a professor at a nearby university, what the main challenges were in teaching the students of today. He stated, "Some just do not want to accept the truth about certain issues," to which I replied, "What issues?" To which he responded, "Well, like global warming for instance and that 97% of the world's scientists believe that man has caused global warming." After I remained silent at his reply for a time, all of those seated at our table turned their heads to hear what I might say. Then he stated, "Surely, Guy, you agree with 97% of the world's scientists on this matter?" I had no ready answer.

While my engineering education in thermodynamics had made me skeptical that man's activities could have such an effect on a large, open-ended thermodynamic system, I was nevertheless at a loss to offer scientific facts to refute the claim. Did 97% of the world's scientists believe in Anthropogenic Global Warming (AGW)? Had I been frittering away my time trying to understand certain topics in advanced physics while the Earth was on a path to thermal destruction? How could I have overlooked what would be arguably one of the most important scientific issues in the history of mankind? Therefore, I set out to research the claim that man-made activities have caused global warming. It has been an interesting journey.

While I have been aware of the concern and controversy over the subject of AGW, I have never been motivated to study it in any detail. Most of my interest of late has been in physics related to the theories on black holes, quantum singularity, and the discovery of the Higgs boson. In my opinion, the detection of the Higgs boson is one of the more important discoveries in physics in the last 100 years (post Einstein's General Relativity Theory in 1915). Interestingly, it has received only a fraction of the attention that has been given to AGW.

In 1972, I received a Bachelor of Science in Mechanical Engineering, with a major in thermodynamics and a minor in thermal fluids. I presumed that my education in mathematics, physics, thermodynamics, and gas dynamics would permit me to understand the underlying scientific principles involved in the fields of climate science related to the issue of AGW. But I was wrong.

During the last three years, I had to study, in some depth, the fields of quantum mechanics, atmospheric physics and spectroscopy, as well as the thermodynamic interactions of the Earth's biosphere: land mass, oceans and atmosphere. I have spent more than two years researching scientific publications on the subject of AGW to understand the various theories that exist concerning the cause(s) of AGW, if indeed it exists. Such knowledge has enabled me to draw certain conclusions about the AGW hypothesis, which I believe to be supported by first principles in science and scientific fact.

Over time, as I studied the issue, I developed what can only be described as a fascination with the subject matter. I was fascinated with the lack of scientific rigor with which the broader climate science community seemed to approach the subject. The AGW hypothesis was accepted as "a given" and the goal of investigations seemed to be to predict the degree to which man-made global warming would occur. That one should question the validity of the hypothesis was deemed heretical, perhaps even immoral.

I was equally fascinated by the "groupthink" propaganda message promulgated by the proponents of the hypothesis. The arguments that "the science was settled" and that "97% of the world's scientists" believed in the AGW hypothesis seemed to me to be a far cry from the discipline and restraint embodied in the scientific method. The more I learned, the more it seemed that proponents of the AGW hypothesis had developed almost a religious fervor. Such phrases as "even if the science is wrong, it is the right thing to do," frequently arose in published debates on the subject.

I have always been interested in the history of the development of scientific knowledge and admired such great minds as Archimedes, Newton, Maxwell, and Einstein. I have always held in high esteem the discipline, logic, objectivity, and methodology of basic scientific research embodied in the scientific method of inquiry. I have always believed to some degree that the pursuit of unbiased, basic scientific research was a noble endeavor, intended to advance man's knowledge of the universe for the ultimate betterment of mankind. It was, if you will, a search for truth.

I suppose what amazed me the most about the subject of AGW was that the "science" I found in support of the hypothesis was far removed from a "noble pursuit." As I began my investigations, I learned that much of the science that supported the AGW hypothesis was not predicated on the scientific method of inquiry. In most cases, the investigator began with the bias that global warming existed. Then, it was presumed that man had caused global warming. The analysis then attempted to validate the hypothesis using empirical or circumstantial evidence in lieu of obtaining data or other facts by experimentation or investigation. This approach is the polar opposite of the scientific method (which we will review in a later chapter), requiring that an effort be made to falsify a hypothesis.

Another interesting aspect of the subject is that many of the proponents of the AGW hypothesis, both scientists and laity, consider the "science to be settled." Such a position appears to be intended to stifle debate or research that would attempt to test or falsify the hypothesis. In addition, the response from proponents of the AGW hypothesis to factual arguments by those who question the hypothesis (deniers) is oftentimes characterized by ad hominem attacks rather than debating the facts underlying the argument. Their purpose is not to debate the factual assertions, but rather to discredit the deniers. This approach is a far cry from Einstein's famous quote: *"No amount of experimentation can ever prove me right; a single experiment can prove me wrong."* [1]

Finally, there is one aspect of the scientific research supporting the AGW hypothesis which I find quite fascinating: Its purpose appears to be to influence public opinion and policy makers with regards to the reduction and/or elimination of man-made emissions of carbon dioxide (CO_2). This objective seems to me to be quite different from the preponderance of scientific inquiry in the past, which focused on advancing scientific knowledge in a particular field. As we will see in a later chapter, the main impetus behind the AGW hypothesis is the United Nations Intergovernmental Panel on Climate Change ("UN IPCC").

In the end, the efforts of the proponents of the AGW hypothesis are intended to convince laity and politicians that the science is settled; that 97% of the world's scientists agree that man has caused global warming; and that immediate steps must be taken to reduce and then eliminate man-made CO_2 emissions into the atmosphere. Their position is that there is no need or time to conduct further scientific research. Immediate action is needed to spare the planet from thermal destruction, if it is not already too late.

In 2016, five percent of the U.S. population were employed as scientists or engineers. [2] In 2015, 24,436 (0.01% of the U.S. population) people received a B.S.M.E. (Bachelor of Science in Mechanical Engineering), widely considered as the most diverse of the engineering disciplines. [3] In 2017, 8,813 students received an undergraduate degree in physics, representing about 0.003% of the U.S. population.[4] The point to be made is that a very small percentage of the population has the technical education to objectively analyze the facts of the matter regarding AGW. Yet, the subject of AGW has become a political one that must ultimately be decided by politicians who are subject to be influenced by public opinion.

The reader will note that there is a minimum of mathematical formulas with which to deal in this book. While I intend to offer an informed opinion on the various technical aspects of the AGW hypothesis, supported by peer-reviewed, published scientific research employing the concept of first principles of science, my target audience is the person who has an interest in the subject of AGW, but perhaps not the technical background to evaluate the scientific claims.

During the course of this work, I hope to provide the reader with a basic knowledge of thermodynamics, the mechanisms of heat transfer, electromagnetic wave theory and the electromagnetic spectrum, in order to assist in understanding the scientific research on the subject. In addition, I will endeavor to explain the pertinent facts concerning quantum physics and spectroscopy in a way that will hopefully be easy to understand.

This knowledge will help the reader evaluate the claims of the proponents of the AGW hypothesis in a rational, factual manner. I submit that common sense, deductive reasoning, an understanding of the technical aspects of the subject matter and the scientific knowledge gained through reviewing published scientific research will suffice in revealing the truth behind the claims of AGW.

CHAPTER THREE

"QUANTIFYING THE CONSENSUS ON ANTHROPOGENIC GLOBAL WARMING IN THE SCIENTIFIC LITERATURE"

IN ANY DISCUSSION with a proponent of the AGW hypothesis, he or she will likely state at some point: "97% of the world's scientists agree that man has caused global warming." It is a mantra that is ubiquitous in the literature supporting the hypothesis, and in the language used by its proponents in debate. This phrase has become the touchstone of the AGW movement. It is often used to truncate debate on the merits of the hypothesis. The obvious implication of the phrase is: "How can you **not** agree with 97% of the world's scientists on the subject matter?" Stated another way, "Do you think that **you** know more than 97% of the world's scientists on this subject?" Therefore, it is important to analyze the origin of this claim in order to evaluate its merit.

"Quantifying the consensus on anthropogenic global warming in the scientific literature (sic)," was published on May 15, 2013, by John Cook, et al., in Environmental Research Letters, Vol. 8, Number 2. Cook refers to the publication as follows: *"This **letter** (emphasis added) was conceived as a 'citizen science' project by volunteers contributing to the Skeptical Science website."*

Skeptical Science is a website founded by Cook in 2007 to promote the anthropogenic global warming hypothesis. In it, Cook describes himself as "a former cartoonist who now researches climate communication at George Mason University." [1] Cook does not purport the "project" to be a scientific study. However, the project report format does follow the basic guidelines used by peer-reviewed publications of scientific articles. I list below the Abstract from this "letter":

"We analyze the evolution of the scientific consensus on anthropogenic global warming in the peer-reviewed scientific literature, examining 11,944 climate abstracts from 1991–2011 matching the topics 'global climate change' or 'global warming.' ***We find that 66.4% of abstracts expressed no position on Anthropogenic Global Warming*** *(emphasis added), 32.6% endorsed Anthropogenic Global Warming, 0.7% rejected Anthropogenic Global Warming and 0.3% were uncertain about the cause of global warming. Among abstracts expressing a position on Anthropogenic Global Warming,* ***97.1% endorsed the consensus position that humans are causing global warming*** *(emphasis added). In a second phase of this study, we invited authors to rate their own papers. Compared to abstract ratings, a smaller percentage of self-rated papers expressed no position on Anthropogenic Global Warming (35.5%). Among self-rated papers expressing a position on Anthropogenic Global Warming, 97.2% endorsed the consensus. For both abstract ratings and authors' self-ratings, the percentage of endorsements among papers expressing a position on Anthropogenic Global Warming marginally increased over time. Our analysis indicates that the number of papers rejecting the consensus on Anthropogenic Global Warming is a vanishingly small proportion of the published research (emphasis added)."* [1]

When I first read the Abstract, I thought that I had misread the author's conclusions based on his own research. Clearly, **the survey results showed that 66.4% of the 11,944 research papers expressed "no position" on arguably the most important topic of "climate science" in history** with which 97% of the world's climate scientists supposedly agreed. How was the 97% claim calculated? Cook disregarded the 66.4% of the research papers that expressed "no position." The 97% claim was calculated using the 33.6% of the abstracts that expressed an opinion on AGW, in the following manner: (32.6)/(32.6+0.7+0.3) = (32.6/33.6) = 97.02%. If reported accurately, Cook's conclusion should have read, "32.6% of the climate scientists surveyed believe in Anthropogenic Global Warming."

Interestingly, in the Results section of the letter, the authors stated the following regarding the second phase of the "study":

"We emailed 8,547 authors an invitation to rate their own papers and ***received 1,200 responses*** *(emphasis added). After excluding papers that were not peer-reviewed, not climate-related or had no abstract, 2,142 papers received self-ratings from 1,189 authors.* ***Among self-rated papers that stated a position on anthropogenic global warming*** *(emphasis added), 97.2% endorsed the consensus. Among self-rated papers not expressing a position on anthropogenic global warming in the abstract, 53.8% were self-rated as endorsing the consensus. Among respondents who authored a paper expressing a view on AGW, 96.4% endorsed the consensus."* [1]

The description of the methodology regarding the second phase of the study begs several questions. First, why would the authors email an invitation to authors to rate papers that were not climate related? Second, if the papers contained no abstract, how could the authors know the subject of the paper and decide to include them in the survey? Finally,

I find it amazing that in the second round of survey requests, only 14% of those surveyed responded to the most important climate science issue in the history of the "science." **86% elected not to respond to the survey.** Based on the manner in which the results of the survey were presented, it is impossible to determine: 1) how many papers stated a position on AGW; 2) how many papers stated no position on AGW; or 3) how many papers stated no position in the Abstract, but were "self-rated as endorsing the consensus," which appears to be no consensus at all! But the authors have an explanation for this aberration. It is listed in the "Discussion" section of the letter:

"Of note is the large proportion of abstracts that state no position on anthropogenic global warming *(emphasis added). This result is expected in consensus situations where scientists…generally focus their discussions on questions that are still disputed or unanswered rather than on matters about which everyone agrees (Oreskes, 2007, p. 72). This explanation is also consistent with a description of consensus as a 'spiral trajectory' in which 'initially intense contestation generates rapid settlement and induces a spiral of new questions' (Shwed and Bearman, 2010);* ***the fundamental science of anthropogenic global warming is no longer controversial among the publishing science community and the remaining debate in the field has moved to other topics*** *(emphasis added). This is supported by the fact that more than half of the self-rated endorsement papers did not express a position on anthropogenic global warming in their abstracts (emphasis added)."* [1]

This logic reminds me of the old quote that is attributed to Yogi Berra, the famous Yankees catcher and later Club manager: *"No one goes there nowadays, it is too crowded."* ~ L.A. Times Sept. 22, 2015.[2]

Since it appears that this "letter" was not subject to scientific peer review, and the authors had an obvious bias, I understand why they would try to skew the results, even though the level of intellectual dishonesty is quite breathtaking. But how is it that intelligent people could be fooled by such chicanery?

Let's read what some of the proponents of AGW have to say on the subject: *"Nobody is interested in solutions if they don't think there's a problem. Given that starting point, I believe it is appropriate to have an over-representation of factual presentations on how dangerous (global warming) is, as a predicate for opening up the audience to listen to what the solutions are..."* ~ Former Vice President, Al Gore (now, chairman and co-founder of Generation Investment Management - a London-based business that sells carbon credits, in an interview with *Grist Magazine,* May 10, 2006, concerning his book, *An Inconvenient Truth.*) [3]

"Each of us has to decide what the right balance is between being effective and being honest." ~ Stephen Schneider (leading advocate of the global warming theory, in an interview for Discover magazine, October 1989) [4]

"Even if the theory of global warming is wrong, we will be doing the right thing — in terms of economic policy and environmental policy." ~ Tim Wirth, former U.S. Senator, Colorado.[5]

"No matter if the science is all phony, there are collateral environmental benefits...Climate change [provides] the greatest chance to bring about justice and equality in the world." ~ Christine Stewart (former Minister of the Environment of Canada quote from the *Calgary Herald,* 1999) [5]

(CNSNews.com) – *"President Obama said Tuesday that he's confident his successor will honor any climate change agreement negotiated in Paris 'because 99.5 percent of scientists and 99 percent of world leaders' think that climate change 'is really important.'"* Dec. 1, 2015

Obama's claim that there is a 99.5 percent consensus among scientists on climate change represents a 2.5 percentage point increase since May 16, 2013. President Obama made this comment one day after the "Quantifying the consensus on anthropogenic global warming in the scientific literature" letter was published, when the President tweeted: "Ninety-seven percent of scientists agree: #climate change is real, man-made and dangerous." [6]

What do a few of the "deniers" have to say about AGW?
"Researchers pound the global-warming drum because they know there is politics and, therefore, money behind it...I've been critical of global warming and am persona non grata." ~ Dr. William Gray (Professor of Atmospheric Sciences at Colorado State University, Fort Collins, Colorado, and leading expert of hurricane prediction in an interview for the "Denver Rocky Mountain News," November 28, 1999) [7]

"Scientists who want to attract attention to themselves, who want to attract great funding to themselves, have to (find a way to) scare the public . . . and this you can achieve only by making things bigger and more dangerous than they really are." ~ Petr Chylek, Professor of Physics and Atmospheric Science, Dalhousie University, Halifax, Nova Scotia, commenting on reports by other researchers that Greenland's glaciers are melting, "Halifax Chronicle-Herald," August 22, 2001) [8]

And perhaps the most telling observation by a climate scientist:

"In the long run, the replacement of the precise and disciplined language of science by the misleading language of litigation and advocacy may be one of the more important sources of damage to society incurred in the current debate over global warming." - Dr. Richard S. Lindzen (leading climate and atmospheric science expert - MIT)[9]

Let's examine each one of these quotes in greater detail to understand the underlying message. Former Vice-President Al Gore:
"Nobody is interested in solutions if they don't think there's a problem. Given that starting point, I believe it is appropriate to have an over-representation of factual presentations on how dangerous (global warming) is, as a predicate for opening up the audience to listen to what the solutions are…" [3] The former Vice-President advocates an "over-representation" of factual presentations on the dangers of global warming, presumably to alarm the "audience" and get them to accept the recommendations for actions to mitigate the dangers.

Dr. Steven Schneider, PhD., (1945-2010) is widely considered to have been the foremost scientist and most outspoken proponent of the AGW hypothesis. Schneider studied engineering at Columbia University, receiving his bachelor's degree in mechanical engineering in 1966. In 1971, he earned a Ph.D. from Columbia in mechanical engineering and plasma physics. Schneider studied the role of greenhouse gases and suspended particulate material on climate as a postdoctoral fellow at NASA's Goddard Institute for Space Studies. He authored or co-authored over 450 scientific papers, conducted research, and published extensively in the field of climatology and global warming.[10] By all accounts, he was a *bona fide* expert in the field. Arguably, he is the most technically qualified scientist advocating for the AGW hypothesis. Let's examine his statement in some detail: *"Each of us has to decide what the* ***right balance is between being effective and being honest*** (emphasis added)." Why can't one be honest and effective on the subject of AGW? The statement suggests that it may be necessary to be dishonest to be effective in conveying the message about AGW.

Timothy Endicott Wirth (born September 22, 1939) is a former United States Senator from Colorado. Wirth, a Democrat, was a member of the House from 1975 to 1987 and was elected to the Senate in 1986, serving one term there before stepping down. Additionally, he served both as Deputy Assistant Secretary for Education for part of the Nixon Administration and later for the Clinton Administration as the first Under Secretary of State for Global Affairs for the U.S. State Department. In the State Department, he worked with Vice President Al Gore on global environmental and population issues, supporting the administration's views on global warming. A supporter of the proposed Kyoto Protocol, Wirth announced the United States' commitment to legally binding limits on greenhouse gas emissions. From 1998 to 2013, he served as the president of the United Nations Foundation, and currently sits on the Foundation's board. [11] Let's examine former Senator Wirth's statement: *"Even if the theory of global warming is wrong, we will be doing the right thing — in terms of economic policy and environmental policy."* Senator Wirth's position is very clear. Even if the science behind the AGW hypothesis is wrong, the issue can be used to address (presumably change) economic policy and environmental policy.

Christine Stewart practiced nursing for a short time before becoming involved in international development work as a volunteer with her husband in Honduras in 1971 to 1972. She served three terms as a Liberal Party Member of Parliament for the Riding of Northumberland in Ontario. During her career, she held the cabinet positions of

Secretary of State for Latin America and Africa, and Minister of the Environment. As the Minister of the Environment, Christine Stewart headed the Canadian delegation to the Kyoto climate change negotiations and signed the Kyoto Accord on behalf of Canada. She pushed for action on the Kyoto Accord, improvements in the Canadian Environmental Protection Act, the Species at Risk Act, and the Canadian Environmental Assessment Act.[12]

*"No matter if the science is all phony, there are collateral environmental benefits...**Climate change [provides] the greatest chance to bring about justice and equality in the world** (emphasis added)."* Mrs. Stewart seems to have the same opinion that Wirth holds on the subject of the AGW hypothesis: regardless of the validity of the hypothesis and the "science" that purportedly underlies it, the end justifies the means. The end is not to attenuate the effects of global warming to prevent an environmental catastrophe, it is to bring about "justice and equality in the world." [13]

I can understand why politicians like Gore, Wirth and Stewart make the comments about using the issue of man-made global warming to accomplish social or economic objectives. Such politicians tend to believe that they know what is best for the ignorant, huddled masses. As such, in their minds the end justifies the means. But what about a man of science like Steven Schneider? Why would he appear to violate the basic tenets of science that underpin the scientific method of inquiry? The answer is found in remarks that he made in an interview with Johnathan Schell in Discover magazine in 1988:

"On the one hand, as scientists we are ethically bound to the scientific method, in effect promising to tell the truth, the whole truth, and nothing

but — which means that we must include all the doubts, the caveats, the ifs, ands, and buts (emphasis added). On the other hand, we are not just scientists, but human beings as well. And like most people we'd like to see the world a better place, which in this context translates into our working to reduce the risk of ***potentially*** *disastrous climatic change (emphasis added). To do that we need to get some broad-based support, to capture the public's imagination. That, of course, entails getting loads of media coverage.* ***So, we have to offer up scary scenarios, make simplified, dramatic statements, and make little mention of any doubts we might have*** *(emphasis added). This* ***'double ethical bind'*** *(emphasis added) we frequently find ourselves in cannot be solved by any formula.* ***Each of us has to decide what the right balance is between being effective and being honest*** *(emphasis added). I hope that means being both."* [4]

The following is a critique of the "Double Ethical Bind in Science" concept by Dr. Judith Curry, PhD., former Professor and Chair of the School of Earth and Atmospheric Sciences at the Georgia Institute of Technology, on the website titled Climate etc. Climate etc. provides the following definition of its purpose: *"...provides a forum for climate researchers, academics and technical experts from other fields, citizen scientists, and the interested public to engage in a discussion on topics related to climate science and the science-policy interface."* [14]

Dr. Curry: *"The double ethical bind arises when a scientist tries to influence the public and policy (emphasis added). It does not arise when a scientist interacts with the media to discuss their latest research finding.* ***This is why advocacy by scientists presents problems both for the scientist and for society*** *(emphasis added)."* [14]

I believe that Dr. Curry has captured the essence of the problem when a scientist-advocate gets involved in the debate about man-made global warming. The problem arises when a scientist tries to influence the public and public policy on an issue for which the science is not settled; that is, the facts are not fully known and undisputed. When it becomes apparent that a scientist-advocate has taken a position on an issue while admitting that the science is not settled, he or she runs the risk of losing credibility in the debate on the matter.

Stanford University maintains a website which contains several articles that Dr. Schneider wrote during his lifetime (he died in July 2010). Interestingly, one of those articles entitled, "Mediarology," was Chapter 7 of his book, *Global Warming*, published in 1989. [4]

"In my view, the best safeguard for public participation in science-based policy issues is to leave subjective probability assessment to the larger scientific community rather than a few charismatic individuals. Some will say, as I noted above, that it is impossible for an expert to maintain his/her scientific objectivity in a value-laden public debate, but after thirty years of striving to do just that, ***I think that science-advocacy can be done honestly*** *(emphasis added)."* [13]

It would seem that Dr. Schneider is of two mindsets on this subject. On the one hand, he states that an individual must "decide what the right balance is between being effective and being honest." On the other, he states that, "science-advocacy can be done honestly." I would imagine the difficulty would arise when trying to decide which dictum to follow.

Interestingly, the current discussion on AGW is not the first time that Dr. Schneider has entered the world of public debate on climate policy and governmental intervention. In 1971, Schneider was second author on a science paper with S. Ichtiaque Rasool entitled, *Atmospheric Carbon Dioxide and Aerosols: Effects of Large Increases on Global Climate.* This paper used a "one-dimensional radiative transfer model to examine the competing effects of cooling from aerosols and warming from CO_2." [15]

The paper concluded that: *"It is* ***projected*** *(emphasis added) that man's potential to pollute will increase six to eightfold in the next 50 years.* ***If this increased rate of injection of particulate matter in the atmosphere should raise the present background opacity by a factor of four, our calculations suggest a decrease in global temperature by as much as 3.5 K*** *(which is a decrease of 6.3°F) (emphasis added). Such a large decrease in the average temperature of Earth, sustained over a period of a few years,* ***is believed to be sufficient to trigger an ice age*** *(emphasis added). However, by that time, nuclear power may have largely replaced fossil fuels as a means of energy production."* It should be noted that this prediction was for the year 2020. Since that pronouncement, there has not been an ice age.

The story made headlines in the *New York Times.* Shortly afterwards, Schneider became aware that he had overestimated the cooling effect of aerosols and underestimated the warming effect of CO_2 by a factor of about three. He had mistakenly assumed that measurements of air particles he had taken near the source of pollution applied worldwide (a very curious error for a scientist of his training). He also found that much of the effect was due to natural aerosols which would not be affected by human activities, so the cooling effect of changes in industrial pollution would be much less than he had calculated (a very basic error in logic). Having found that recalculation showed that global warming was the more likely outcome, he published a retraction of his earlier findings in 1974.[15]

So, what is one to make of Dr. Schneider's very basic errors in his research assumptions and his 180-degree pivot on global cooling to global warming? Do the vary basic errors in his earlier calculations damage his credibility? Does his departure from the traditional role of a scientist as practitioner of the scientific method to one who advocates public policy discredit his opinions? Perhaps a study in contrast with the other scientist who is widely recognized as an eminent authority in the field of climate science, and an opponent of the AGW hypothesis, may shed some light on these questions and assist us with the answers.

Dr. Richard Lindzen, PhD., (born February 8, 1940) is an American atmospheric physicist known for his work in the dynamics of the middle atmosphere, atmospheric tides and ozone photochemistry. Lindzen attended the Bronx High School of Science (winning Regents' and National Merit Scholarships), Rensselaer Polytechnic Institute and Harvard University. He received a B.A. in physics from Harvard in 1960, followed by a M.S. in applied mathematics in 1961 and a Ph.D. in applied mathematics in 1964. His doctoral thesis, "Radiative and Photochemical Processes in Strato and Mesospheric Dynamics," addressed the interactions of ozone photochemistry, radiative transfer, and dynamics in the middle atmosphere. He has published more than 200 scientific papers and books. He was a lead author of Chapter 7, "Physical Climate Processes and Feedbacks," of the Intergovernmental Panel on Climate Change's Third Assessment Report for climate change. He has criticized the scientific consensus about climate change and what he has called "climate alarmism." He was Alfred P. Sloan Professor of Meteorology at MIT from 1983 until his retirement, which was reported in the Spring 2013 newsletter of MIT's Department of Earth, Atmospheric and Planetary Sciences (EAPS). On December 27, 2013, the Cato Institute announced that Dr. Lindzen had been appointed a Distinguished Senior Fellow in their Center for the Study of Science. [16]

Unlike Schneider, Lindzen has written only two books, both highly technical in nature. Both resulted from basic research that Lindzen had conducted in association with atmospheric physics. One, *Dynamics in Atmospheric Physics,* published in 2005, is a textbook for graduate students on the structure and dynamics of the atmosphere. The second, *Atmospheric Tides,* published in 1970, deals with the thermal and gravitational tides that affect temperature. [16] Neither book deals with the subject of global climate change. During his career, Lindzen has taken exception with the theories of some climate scientists with respect to the role of water vapor in the atmosphere, the effect of increasing atmospheric concentrations of CO_2 on global warming, and the efficacy of the mathematical models used by the UN IPCC to predict temperature.

So, how does Lindzen's view of the issues involved with the AGW hypothesis differ from those of Schneider? I will let Lindzen answer that question in his own words as contained in a letter that he wrote to President Donald Trump, dated March 9, 2017: [17]

"The UNFCCC (United Nations Framework Convention on Climate Change) was established twenty-five years ago to find scientific support for dangers from increasing carbon dioxide. While this has led to generous and rapidly increased support for the field, ***the purported dangers remain hypothetical, model-based projections*** *(emphasis added). By contrast, the benefits of increasing CO_2 and modest warming are clearer than ever, and they are supported by dramatic satellite images of a greening earth."*

"The UN's Intergovernmental Panel on Climate Change (IPCC) no longer claims a greater likelihood of significant as opposed to negligible future warming. It has long been acknowledged by the IPCC that climate change prior to the 1960's could not have been due to anthropogenic greenhouse gases. Yet, pre-1960 instrumentally observed temperatures show many warming episodes, similar to the one since 1960, for example, from 1915 to 1950, and from 1850 to 1890. None of these could have been caused by an increase in atmospheric CO_2. ***Model projections of warming during recent decades have greatly exceeded what has been observed*** *(emphasis added). The modeling community has openly acknowledged that the ability of existing models to simulate past climates is due to numerous arbitrary tuning adjustments."*

"Observations show no statistically valid trends in flooding or drought, and no meaningful acceleration whatsoever of pre-existing long-term sea-level rise (about six inches per century) worldwide. Current carbon dioxide levels, around 400 parts per million are still very small compared to the averages over geological history, when thousands of parts per million prevailed, and when life flourished on land and in the oceans. Calls to limit carbon dioxide emissions are even less persuasive today than 25 years ago. ***Future research should focus on dispassionate, high-quality climate science*** *(emphasis added), not on efforts to prop up an increasingly frayed narrative of 'carbon pollution.'* ***Until scientific research is unfettered from the constraints of the policy-driven UNFCCC*** *(emphasis added), the research community will fail in its obligation to the public that pays the bills."*

I think that Lindzen's approach to the issue is notable in its difference from Schneider's approach in two fundamental aspects. First, Lindzen makes the point that the science is "not settled." He points out the contradictions between the observable phenomena with respect to the Earth's climate and that which was predicted from the model-based projections. Second, he calls for future research to obtain factual information in lieu of the "hypothetical, model-based projections." In essence, Lindzen is indicting the lack of scientific facts to support the AGW hypothesis and calling for more objective, high quality scientific research to inform debate on the issue, and a return to the scientific method of inquiry. This sounds to me like a very reasonable view. Lindzen is not trying to truncate debate; he is proposing that it should be an informed debate based on scientific evidence.

Role of the Scientist as Advocate

The role of the scientist as advocate for the benefit of society is not unprecedented. However, it has generally been restricted to issues in science where the facts are well known and undisputed. On August 2, 1939, the Einstein–Szilárd letter was written by Leó Szilárd, signed by Albert Einstein and sent to United States President Franklin D. Roosevelt. Written by Szilárd in consultation with fellow Hungarian physicists Edward Teller and Eugene Wigner, the letter warned that Germany might develop atomic bombs and suggested that the United States should start its own nuclear research program. [18] The letter prompted action by Roosevelt to initiate a research effort on the development of nuclear weapons. On December 2, 1942, the world's first self-sustaining, controlled nuclear chain reaction took place under the direction of Enrico Fermi, a Nobel Prize-winning scientist, at the University of Chicago. Fermi's work paved the way for a variety of advancements in nuclear science, including the Manhattan Project, which developed the first atomic bombs employed by the U.S. military to end World War II.

Einstein realized the threat of the terrible force that he had helped to unleash. Because Einstein felt responsible for the development of the atomic bomb, he was compelled to advise against its further development and use. In the *Russell-Einstein Manifesto* issued in London, England on July 9, 1955, Einstein argued against the development and proliferation of nuclear weapons. [19]

The challenge with the subject of AGW is that the science is not settled, regardless of what the proponents might argue. However, it seems that the issue of AGW is to be adjudicated in the court of public opinion by persons who do not possess all the relevant facts or the technical knowledge to evaluate the facts with respect to the various claims. Therefore, it is so important that the American public, politicians, and policy makers, be informed as to the flaws in the science involved in the AGW hypothesis. In my opinion, the impact of the unnecessary reduction or elimination of the use of fossil fuels would have a significant adverse effect on the standard of living of every person in both the developed and developing countries.

As we shall learn in later chapters in this book, the science is far from "settled or proven," with respect to the global warming hypothesis. The thermodynamic interactions which take place among the various systems of the Earth's biosphere, are complex and depend on numerous variables that are constantly changing and difficult to quantify. Why then would Dr. Schneider and others advocate for action to address a problem that has not been fully understood, or perhaps even proven to exist? It has to do with what these advocates believe is their social duty to society. It has to do with the belief that if the hypothesis is correct, then in their judgment, the impending consequences to society compel immediate action (in the case of global warming) to save the planet and mankind. However, if the hypothesis is later proven to be wrong, then (in the scientist advocate's judgment) the efforts undertaken were beneficial to mankind regardless, because economic and social inequalities had been redressed.

CHAPTER FOUR

WHAT DO THE WORLD'S SCIENTISTS BELIEVE ABOUT MAN-MADE GLOBAL WARMING?

I FIRST BEGAN my investigations into the facts about AGW in June 2017. To test the claim that 97% of the world's scientists endorsed the man-made global warming "consensus," I set about to obtain the views of some of the world's universities and scientific organizations on the subject. Accordingly, I visited each of their websites in October of 2017. Below, I have listed their views about the subject matter as published on their websites in 2017, along with my commentary at the time. In addition, I have also listed their views on the subject matter as of October 2019, along with my comments. In several cases, their change of view on the subject in such a short time was striking.

June 2017: Germany - The Helmholtz Association

Germany: *"The Helmholtz Association pursues the long-term research goals of state and society to maintain and improve the livelihoods of the population. To do this, the Helmholtz Association conducts top-level research to identify and explore the major challenges facing society, science and the economy."* [1]

"The effects of global climate change on particular regions ***vary significantly*** *(emphasis added). Farmers, coastal engineers, town planners and other* ***decision makers need first-hand information on regional climate change in order to adapt their region to the effects*** *(emphasis added)." [1]*

"The Helmholtz Association has therefore decided to initiate a German network of regional climate offices. Each of the four Regional Helmholtz Climate Offices is focusing on a certain region. ***We integrate regional climate change information based on the latest research projects and make scientific results understandable to the public*** *(emphasis added)."* [1]

A very sensible, scientific, and pragmatic approach based on a plan of scientific investigation, the Helmholtz Association is studying the effect that climate change might have on various regions of the country. They further state that remediation plans will be considered, as necessary.

October 2019: Germany - The Helmholtz Association

"Forests are drying out in Brandenburg, glaciers are melting in Bavaria, hurricanes are raging in the Caribbean, and the Arctic is losing enormous amounts of ice. Extreme events such as these are constantly on the rise and have one clear cause: ***climate change*** *(emphasis added). Humanity now faces a challenge that is as much self-made as it is inconceivably immense.* ***It must curb the causes of climate change as quickly and sustainably as possible*** *while at the same time finding ways to adapt to a massively changing world (emphasis added)."* [1]

"Experts from all over the world have been warning of the need for rapid

action for years. Only last year a special report by the Intergovernmental Panel on Climate Change (IPCC) was published and U.N. climate change conferences are held annually. At the 2015 United Nations Climate Change Conference in Paris, the international community agreed to limit global warming to +1.5 degrees Celsius if possible and to a maximum of +2 degrees Celsius compared to preindustrial levels. The numbers are clear: ***Climate change is real and is predominantly caused by us humans*** *(emphasis added)." [1]*

It seems that the scientists at the Helmholtz Association work very quickly! It took them less than two years to abandon their well-reasoned research plan to conclude that global warming was a real, worldwide problem and not a local phenomenon. You will note that they invoke the 2015 publication of the IPCC in their remarks to substantiate their position. It should be noted that the 2015 UN IPCC report was published two years before the June 2017 comments on their website, leaving one to question the real reason for such a curious change of mind on the subject.

2017: England - The British Royal Society

England: The British Royal Society seemed to be a bit schizophrenic in their view(s) on AGW in 2017; their website expressed two opinions: [2] *"Rigorous analysis of all data and lines of evidence shows that most of the observed global warming over the past 50 years or so cannot be explained by natural causes and instead* ***requires a significant role for the influence of human activities*** *(emphasis added)."*

"In order to discern the human influence on climate, scientists must consider many natural variations that affect temperature, precipitation, and other aspects of climate from local to global scale, on

mescales from days to decades and longer** (emphasis added). One natural variation is the El Niño Southern Oscillation (ENSO), an irregular alternation between warming and cooling (lasting about two to seven years) in the equatorial Pacific Ocean that causes significant year-to-year regional and global shifts in temperature and rainfall patterns. Volcanic eruptions also alter climate, in part increasing the amount of small (aerosol) particles in the stratosphere that reflect or absorb sunlight, leading to a short-term surface cooling lasting typically about two to three years.* ***Over hundreds of thousands of years, slow, recurring variations in Earth's orbit around the Sun, which alter the distribution of solar energy received by Earth, *(emphasis added) have been enough to trigger the ice age cycles of the past 800,000 years."*

One is left to wonder what exactly the British Society intends to do to reach a conclusion on its views regarding AGW. The paragraph above suggests that British scientists should conduct studies on the "natural variations" of the Earth's climate system to differentiate the effect of human influences on the climate versus natural causation. A very logical, pragmatic view.

2019: England - The British Royal Society

"The Intergovernmental Panel on Climate Change (IPCC) special report on the impacts of global warming of 1.5°C outlines the advantages of limiting the rise in average global temperature to 1.5C above pre-industrial levels, and the pathways needed to achieve this. Whilst this requires major and widespread action within the next decade, it could avoid many damaging impacts. This briefing provides a summary of the IPCC's findings and what these mean for the UK, identifying what UK policymakers can do now, both in terms of UK policy and globally, to enable the UK to play its role in limiting warming to as close as possible to 1.5°C." [2]

Like the Germans in 2019, within a two-year period, the British seem to have accepted man-made global warming as a fait accompli and the most recent IPCC report as the state of the art on the subject. There is no more discussion about an effort to differentiate the effect of human influence from natural causation.

2017: Cambridge University

Cambridge University has developed "Research Themes" listed on their website: [3]

"Atmosphere-Biosphere Interactions; Ocean and sea ice dynamics; Cryosphere and Sea-Level Rise, Earth System Modeling and Past Climate Change." Undoubtedly one of the most scholarly approaches to the subject listed online. Research is not focused on proving that AGW has occurred with the presumption that man is responsible, but rather, developing the analytical and research tools to be able to determine interactions within the eco system, and measure what impacts have occurred as well as a model to predict the future. Very restrained, very British.

2019: Cambridge University

Cambridge has changed the design of their website somewhat since 2017. The site no longer contains general statements about the subject of AGW; but rather, focuses on the student and faculty research on a number of subjects. AGW does not appear to be one of them; or at least not one that was discussed.

2017: Oxford University

Oxford seems to take a restrained position similar to Cambridge:[4]

"The decision on whether to increase the ambition of climate change mitigation efforts to stabilize temperatures at 1.5°C rather than 2°C above pre-industrial (the goal of the Paris Climate Accord) is arguably one of the

most momentous to be made in the coming decade, and is ***currently poorly served by the paucity of scientific analysis of the relative risks associated with these two outcomes*** *(emphasis added) (James et al, 2016),* ***particularly regarding the role of extreme weather*** *(emphasis added). The Conference of the Parties of the United Nations Framework Convention on Climate Change (UNFC-CC), in its Paris Agreement of 2015, invited the Intergovernmental Panel on Climate Change (IPCC) to prepare a Special Report in 2018 "on the impacts of global warming of 1.5°C above pre-industrial levels and related greenhouse gas emission pathways.* ***To inform such an assessment, research will need to be undertaken immediately, over the period 2016 to 2017."***

As opposed to just accepting the findings of the IPCC, Oxford states that "there is a paucity of scientific analysis" and "research will need to be undertaken immediately" to make an informed decision. However, given the scope of the proposed research, one would wonder if a two-year period would be sufficient to conduct that research. It should be noted that this is the first mention of the concept of evaluating what impact climate changes might have should they occur.

2019: Oxford University

Like Cambridge, Oxford seems to have moved away from a comprehensive statement by the University such as above. The research section of the website lists many research projects by students and faculty on the subject matter. However, it is clear that the tenor of the description of those research projects accepts the AGW hypothesis and focuses on an effort to evaluate the benefits of proposed mitigative actions.

2017: France – French Academy of Sciences

"The IPCC's scientific conclusions are unquestionable: there can be no

doubt that global warming is taking place and the IPCC considers it extremely likely (more than 95% probability, compared to 90% in 2007 and 66% in 2001) that human activity is responsible for the increase in mean global temperatures since the mid 20th century." [5]

"At European Union (EU) level, France advocates an ambitious position, with the goal of reducing greenhouse gas emissions by 40% by 2030, then 60% by 2040 (emphasis added) (compared with 1990), to ensure the EU maintains an ambitious and exemplary approach." The French are all in for the global warming hypothesis. [5]

And there is more from newly elected French President Emanuel Macron. He offered climate scientists in the United States around 1.5 M Euros each to emigrate to France to study climate change. "I do know how your new president now has decided to jeopardize your budget, your initiatives, and he is extremely skeptical about climate change," he said. "I have no doubt about climate change." Macron went on to promise robust funding for climate initiatives. [5]

2019: France - Academy of Sciences

Interestingly, a search of the Academy's website does not show any research projects or other discussions about the subject of AGW. I must assume that it is considered "settled science" in France and that at least 97% of the French climate scientists agree.

2017: The U.S. National Academy of Sciences

WASHINGTON – *"The U.S. National Academy of Sciences and the Royal Society, the national science academy of the U.K., released a joint publication today (Feb. 27, 2014) in Washington, D.C.,* ***that explains the clear evidence that humans are causing the climate to change*** *(emphasis added), and that addresses a variety of other key questions commonly asked about climate change science."* [6]

"As two of the world's leading scientific bodies, we feel a responsibility to evaluate and explain what is known about climate change, at least the physical side of it, to concerned citizens, educators, decision makers and leaders, and to advance public dialogue about how to respond to the threats of climate change," said NAS President Ralph J. Cicerone.

It should be noted that the joint statement was not based on research done separately by either organization, but the findings of the IPCC 2014 Report referenced earlier. Finally, I think that it is curious that the British Royal Society would issue the statement on their website in June 2017 calling for research in an effort to distinguish natural causation from man-made activities as relates to the subject of global warming **after** issuing this joint statement with the U. S. National Academy of Sciences.[6]

2019: The U.S. National Academy of Sciences
"Recently, questions have been raised about climate science *(emphasis added). The National Academies have addressed many of these questions in our independent, evidence-based reports. We are speaking out to support the cumulative scientific evidence for climate change and the scientists who continue to advance our understanding."* [7]

"Scientists have known for some time, from multiple lines of evidence, that humans are changing Earth's climate, primarily through greenhouse gas emissions *(emphasis added). The evidence on the impacts of climate change is also clear and growing. The atmosphere and the Earth's oceans are warming, the magnitude and frequency of certain extreme events* *are increasing, and sea level is rising along our coasts."* [7]

"The National Academies are focused on further understanding climate change and how to limit its magnitude and adapt to its impacts, including on health. We also recognize the need to communicate more clearly what we know. To that end, in 2018, the National Academies launched an initiative to make it easier for decision makers and the public to use our extensive body of work to inform their decisions. In addition, we will be expanding our Basic Science communications effort to include clear, concise, and evidence-based answers to frequently asked questions about climate change." [7]

"A solid foundation of scientific evidence on climate change exists *(emphasis added). It should be recognized, built upon,* ***and most importantly, acted upon*** *for the benefit of society (emphasis added)."* [7]

I find it noteworthy that the National Academies felt the need to take up a defense of climate science and the conclusion that "humans are changing the Earth's climate." While they state that they have "addressed many of those questions," they seem to believe that they need to reiterate their view that scientific evidence supports the existence of climate change. Given the National Academies' admission that "questions have been raised about climate science," it makes one wonder why there is such a sense of urgency to act. To paraphrase Queen Gertrude from Shakespeare's play *Hamlet,* "the National Academies doth protest too much, me thinks!"

2017: The National Aeronautics and Space Administration (NASA)
"Multiple studies published in peer-reviewed scientific journals show that 97 percent or more of actively publishing climate scientists agree *(emphasis added): Climate-warming trends over the past cen-*

tury are extremely likely due to human activities. In addition, most of the leading scientific organizations worldwide have issued public statements endorsing this position. The following is a partial list of these organizations, along with links to their published statements and a selection of related resources." [8]

While I would be disappointed, I suppose that I could understand if NASA accepted the global warming hypothesis and referenced the U.S. Academies position or even the UN IPCC statements, given President Obama's stance on AGW and the issue of funding for NASA. However, I find it incredible that their first citation to validate their position was the "survey" done by the "citizen scientists," led by John Cook, et. al., for the *Skeptical Science* website. Multiple studies published in peer-reviewed scientific journals? Not even Cook et. al. claim that their work was peer reviewed.

I suppose that of all the world's scientific organizations, I would have expected NASA to be the most discerning in its deliberations concerning the acceptance of the global warming hypothesis. How is it possible that the largest, most visible scientific organization in the U.S. (and at one time, perhaps the world), that has historically typified excellence in applied physics and engineering and landed a manned rocket on the moon on July 20, 1969, has fallen so far? Is funding so important that they simply fall in line with the "consensus"?

2019: NASA

NASA does not appear to have updated its website with any current information concerning the subject of AGW.

2017: Russian Academy of Sciences

"A new scientific paper authored by seven scientists affiliated with the Russian Academy of Sciences was just published in the scientific journal 'Bulletin of the Russian Academy of Sciences: Physics.' The scientists dismiss both "greenhouse gases" and variations in the Sun's irradiance as significant climate drivers, and instead embrace cloud cover variations — modulated by cosmic ray flux — as a dominant contributor to climate change."

Cosmic rays are high-energy protons and atomic nuclei which move through space at nearly the speed of light. They originate from the sun, from outside of the solar system, and from distant galaxies. Upon impact with the Earth's atmosphere, cosmic rays can produce showers of secondary particles that sometimes reach the surface. *"As cosmic ray flux increases, more clouds are formed on a global scale. More global-scale cloud cover means more solar radiation is correspondingly blocked from reaching the Earth's surface (oceans). With an increase in global cloud cover projected for the coming decades (using trend analysis),* ***a global cooling is predicted*** *(emphasis added)." [9,10]*

At least the Russian scientists performed their own research. They maintain that global cooling will result by 2060 (about a 0.7°C decline) due to an increase in "cosmic ray flux," which would increase cloud cover, thereby reducing the amount of solar irradiation that reaches the Earth's surface. [10]

The Russian Academy has not published any additional statements concerning the subject of AGW on their website. However, in a meeting in Paris in August 2019, President Vladimir Putin discussed climate change with French President Emmanuel Macron, saying Moscow is adopting programs and allocating major resources in combating it.

"We take this matter very seriously," Mr. Putin said. "We have to coordinate our efforts. We are prepared for this joint work." Mr. Putin did not detail steps that Russia would take to deal with AGW, nor did he reference the Russian Academy of Sciences. However, I would be surprised if any scientists in the Russian Academy publicly disagreed with Mr. Putin. [11]

2017: Chinese Academy of Sciences - The Institute of Atmospheric Physics "New Study Reveals the Atmospheric Footprint of the Global Warming Hiatus."

"The increasing rate of the global mean surface temperature was reduced from 1998 to 2013, known as the global warming hiatus or pause *(emphasis added). Great efforts have been devoted to the understanding of the cause. The proposed mechanisms include the internal variability of the coupled ocean-atmosphere system, the ocean heat uptake and redistribution, among many others. However, the atmospheric footprint of the recent warming hiatus has been less concerned. Both the dynamical and physical processes remain unclear."* In essence, this statement says that: global warming of the Earth's surface was paused from 1998 to 2013; and the Chinese have undertaken great efforts to understand what caused the pause in global warming. The proposed causes of the pause might include intrinsic variations in the heat transfer mechanisms between the ocean and atmosphere and the manner in which the planet's oceans absorb and redistribute heat. Finally, the Chinese appear to state that they don't know which processes account for the pause in global warming. [12] In addition, the findings of an additional Chinese Academy of Sciences research paper regarding the troposphere were summarized:

"In a recent paper (2017) published in Scientific Report, Bo Liu and Tianjun Zhou from the Institute of Atmospheric Physics, Chinese Academy of Sciences, have investigated the atmospheric anomalous features during the global warming hiatus period (1998-2013).[13] ***They show evidence that the global mean tropospheric temperature also experienced a hiatus or pause*** *(emphasis added). To understand the physical processes that dominate the warming hiatus, they decomposed the total temperature trends into components due to processes related to surface albedo (reflection of solar irradiance from the Earth's surface), water vapor, cloud, surface turbulent fluxes and atmospheric dynamics. The results demonstrated that the hiatus of near surface temperature warming trend is dominated by the decreasing surface latent heat flux compared with the preceding warming period, while the hiatus of upper tropospheric temperature is dominated by the cloud-related processes. Further analysis indicated that atmospheric dynamics are coupled with surface turbulent heat fluxes over lower troposphere and coupled with cloud processes over upper troposphere."*

In this statement, the Chinese identify two prospective causes for the global warming hiatus in the troposphere. First, a decrease in surface latent heat flux (the amount of heat transferred from the Earth's surface to the near surface air by evaporation) caused a reduction in the temperature of the air close to the surface of the Earth. As a result, the cooler air near the surface did not transfer as much heat to the troposphere, when compared to prior periods. Second, they identify "cloud-related processes" as the cause of a reduction in the temperature of the upper troposphere. Presumably, the cloud-related processes resulted in more cloud cover, which blocked sunlight from reaching the upper troposphere, causing cooling. Finally, they state that "atmospheric dynamics" (changes in the characteristics of the atmosphere) are a function of changes in surface turbulent heat flux which affects the temperature of the lower troposphere and cloud processes which affect the temperature of the upper troposphere.

2019: Chinese Academy of Sciences - The Institute of Atmospheric Physics

Interestingly, the website for the Chinese Academy of Sciences contains many articles that deal with research about global warming, but no general statements on the subject *per se*. Most statements address the contradictions between the increase in CO_2 concentration in the atmosphere and the global warming "hiatus." Several deal with the subject of cooling of the oceans in certain parts of Asia and the connection between the thermodynamic operations of the ocean/atmosphere and the climate. It seems that the Chinese still do not accept the AGW hypothesis and are focused on research to try to understand what drives climate and weather changes. [14]

In conclusion, it appears that during the period 2017 to 2019, many of the world's scientific and academic societies abandoned the view that more research was needed on the subject of man's influence on the climate and the issue of AGW. It is difficult to imagine that sufficient scientific research could have been conducted by the members of these organizations on the numerous topics identified as pertinent to the issue of AGW in such a short period of time. Instead, they appear to have adopted the UN IPCC view of the cause of AGW. The obvious question is: what caused them to change their opinions in such short order? We will review the possible reasons for this change of mind in more detail in Chapter 12.

CHAPTER FIVE

THE UNITED NATIONS' INTERGOVERNMENTAL PANEL ON CLIMATE CHANGE

PROPONENTS OF AGW often cite publications by the Intergovernmental Panel on Climate Change (IPCC) as their source for scientific confirmation of the AGW hypothesis. I list below the IPCC's definition of its mission:

"The Intergovernmental Panel on Climate Change (IPCC) is the leading international body for the assessment of climate change. It was established by the United Nations Environment Programme (UNEP) and the World Meteorological Organization (WMO) in 1988 ***to provide the world with a clear scientific view on the current state of knowledge in climate change and its potential environmental and socio-economic impacts*** *(emphasis added)."* [1]

U.S. taxpayers have funded over half of the IPCC budget since 2000 at $31.1 million. The U.S. State Department provided $19 million, while the U.S. Global Change Research Program funded $12.1 million. [2]

On its website, the IPCC lists the following under the "Reports" section: *"The* ***main activity of the IPCC is to provide*** *at regular intervals*

Assessment Reports of the state of knowledge on climate change *(emphasis added)."* [2] It would stand to reason that the main purpose of the UN IPCC would be to report on the state of the knowledge within the world scientific community about climate change. However, as we will see, the main purpose of the UN IPCC seemed to quickly evolve over the years.

In the seminal work of the IPCC in 1990, the major conclusion from the study was [3]:
"Under the IPCC Business-as-Usual (Scenario A) emissions of greenhouse gases, a rate of increase of global mean temperature during the next century of about 0.3°C per decade (with an uncertainty range of 0.2°C to 0.5°C per decade), ***this is greater than that seen over the past 10,000 years.*** *(sic) (emphasis added) This will result in a likely increase in global mean temperature of about 1°C (3.8°F) above the present value by 2025 and 3°C (5.4°F) before the end of the next century. The rise will not be steady because of the influence of other factors."* There was no accompanying statement about the prospective impact on the environment. **It should be noted that the uncertainty range of the estimate is equal to the value of the projected increase.** Based on the data obtained from the NOAA satellite measurements as contained in the UAH temperature dataset, the average increase in the temperature anomaly of the lower troposphere for the period 1990 to 2020 was 0.402°C (0.134 °C /decade), about 40% of that predicted by the IPCC in 1990.

The IPCC published its AR5 report in 2014. Under the section entitled, "Headline Statements from the Summary for Policymakers "Observed Changes in the Climate System," it stated the following [4]:
"Warming of the climate system is unequivocal, and since the 1950s, many of the observed changes are unprecedented over decades to millennia. The

atmosphere and ocean have warmed, the amounts of snow and ice have diminished, sea level has risen, and the concentrations of greenhouse gases have increased. ***Each of the last three decades has been successively warmer at the Earth's surface than any preceding decade since 1850. In the Northern Hemisphere, 1983–2012 was likely the warmest 30-year period of the last 1400 years (medium confidence)*** *(emphasis added)."*

First, it should be noted that the IPCC has stated in a different publication that an accurate historical temperature database did not exist prior to the advent of satellite measurements using microwave sensing devices to accumulate data on the temperature of the lower troposphere. Second, it was during the period 1998 to 2013 that land surface temperature datasets recorded a <u>decrease</u> in the temperature anomaly, often referred to as the "global warming hiatus or pause," as the Chinese Academy of Science noted on its website. [5] It is curious that the world body appointed by the U.N. to provide regular reports on the state of the world knowledge of climate science would seemingly be unaware of the global warming pause.

Clearly, the AR5 report in 2014 contained a dire prediction of things to come for Earth's climate. Perhaps the most damning statement in the report: [4]
"Human influence has been detected in warming of the atmosphere and the ocean, in changes in the global water cycle, in reductions in snow and ice, in global mean sea level rise, and in changes in some climate extremes." This evidence for human influence has grown since AR4. ***It is extremely likely that human influence has been the dominant cause of the observed warming since the mid-20th century*** *(emphasis added)."*

It appears that the ardor of the authors of the most recent report seems to have risen in direct proportion to the alleged increase in the global mean temperature over the same period. *For the first time, the writers of the report cited the impacts of global warming to include "changes in the global water cycle, in reductions in snow and ice, and in changes in some climate extremes." [4]* ***It should be noted that none of the aforementioned can be verified by objective measurement.*** The report offers no insight as to how these parameters could or might have been calculated or what effect they might have on life on Earth.

I think that these statements demonstrate several very important aspects of the *modus operandi* of the UNIPCC that typifies the way they communicate. First, the IPCC is given to engaging in hyperbole, while at the same time qualifying their statements using such terns as "detected and "extremely likely", which is not the stock in trade of objective scientists. Second, the statement references parameters which cannot be objectively measured such as "changes in the global water cycle", the amount of "snow and ice" and "changes in some climate extremes." This is clearly not the precise language demanded of scientific investigations. Finally, to state that "it is extremely likely that human influence has been the dominant cause of the observed warming" indicates a bias not supported by objective facts.

Notwithstanding the fact that the ability of the scientific community engaged in AGW analysis to measure these variables is questionable, I think that the use of such broad language by the IPCC in their reports without fact-based evidence is indicative of the lack of methodological rigor and discipline in the conduct of their analyses. The IPCC has acknowledged that **no accurate historical temperature database going back to the 1950's exists with respect to the temperature of the world's surface, oceans or atmosphere.** [4] The use of such anecdotal "evidence" is indicative of a major weakness in the analytical model that the UN IPCC frequently employs; that is, associating correlation with causation.

In its *Report by Chapters* under the section "Observations: Atmosphere and Surface," the IPCC report states: "*Based on multiple independent analyses of measurements from radiosondes and satellite sensors,* ***it is virtually certain*** *(emphasis added) that globally the troposphere has warmed, and the stratosphere has cooled since the mid-20th century.*" [6]

Again, it should be noted that reliable temperature data is not available prior to 1979. Publicly available data from NOAA satellite measurements from 1979 to present shows a high degree of variability in the temperature anomalies of the lower troposphere with a *de minimis* increase in temperature over the 40-year period. [7] In addition, there is no long-term temperature database for the stratosphere.

The UN IPCC has been the driving force behind the claims of global warming since its inception. Therefore, I think that it would be useful to review the IPCC published statements on the history of the activities of the organization to understand how their thinking has evolved on the issue:

"Since 1988, the IPCC has had five assessment cycles and delivered five Assessment Reports, the most comprehensive scientific reports about climate change produced worldwide *(emphasis added). It has also produced a range of Methodology Reports, Special Reports and Technical Papers, in response to requests for information on specific scientific and technical matters from the United Nations Framework Convention on Climate Change (UNFCCC), governments and international organizations.*" [1]

"In 1990, the First IPCC Assessment Report (FAR) underlined the importance of climate change as a challenge with global consequences *(emphasis added) and requiring international cooperation. It played a decisive role in the creation of the UNFCCC,* ***the key international treaty to reduce global warming and cope with the consequences of climate change*** *(emphasis added).[3] The Second Assessment Report (SAR) (1995) provided important material for governments to draw from in the run-up to adoption of the Kyoto Protocol in 1997.[8] The Third Assessment Report (TAR) (2001) focused attention on the impacts of climate change and the need for adaptation.[9] The Fourth Assessment Report (AR4) (2007) laid the groundwork for a post-Kyoto agreement,* ***focusing on limiting warming to 2°C*** *(emphasis added).[10] The Fifth Assessment Report (AR5) was finalized between 2013 and 2014. It provided the scientific input into the Paris Agreement." [11].* The "key international treaty" to which the statement above refers is the Kyoto Protocol. It should be noted that during the Kyoto Protocol, world-wide emissions of CO_2 increased by 32%. By any measure (aside from the increase in trading of carbon credits), the Kyoto Protocol was a dismal failure.

"The IPCC is currently in its Sixth Assessment cycle where it will prepare three Special Reports, a Methodology Report and the Sixth Assessment Report. The first of these Special Reports, Global Warming of 1.5°C (SR15), was requested by world governments under the Paris Agreement. In May 2019, the IPCC finalized the 2019 Refinement – an update to the 2006 IPCC Guidelines on National Greenhouse Gas Inventories. The Special Report on Climate Change and Land (SRCCL) will be finalized in August 2019 and the Special Report on the Ocean and Cryosphere in a Changing Climate (SROCC) will be finalized in September 2019. The Sixth Assessment Report (AR6) is expected to be finalized in 2022 in time for the first global stock take the following year." [1]

SR 15 was published by the UNIPCC on October 8, 2018. The report stated that there was a "high degree of confidence" that human activities had caused global warming of approximately 1°C. [12] In addition, it stated that limiting global warming to 1.5°C would be possible, but "global net CO_2 emissions would have to fall by 45% from 2010 levels by 2030 and reach net zero by 2050." Further, the report predicted that as a result of the 1.5°C warming, there would be catastrophic impacts on world-wide agriculture involving reduced water supply, reduced crop yields, reduced nutritional value of the food produced, increased spread of disease by farm animals and an increase in "vector-borne diseases" such as malaria and dengue fever. Recommendations from the report included increasing the percentage of primary energy production (electric power generation) involving renewable energy to 60% of the total; increasing the use of electric-powered transportation; reducing coal-fired power generation to 5% of the total; and a larger role for nuclear power generation. In summary, SR 15 forecast a gloomy future for life on Earth with calamity on every front should a 1.5°C temperature increase come to pass; but, predicted even worse conditions should warming increase to 2°C. I find it amazing that anyone could believe that such detailed, dire consequences could be predicted for the Earth's future with such a small, predicted temperature increase **without any objective corroborating data; or, a demonstrated ability to back-test predictive models.** Keep in mind that global mean temperature is an abstract; it is a meaningless value in science and thermodynamic analysis.

On October 8, 2019, the UNIPCC released the "Special Report on Climate Change and Land" ("SRCCL"). It was written by 107 Coordinating Lead Authors, 96 Contributing Authors, all from 52 countries. At 36 pages in length, it addressed every possible aspect of land use.

The report stated that it was "very likely" that human influence was responsible for warming of land surface air temperature of 1.53°C during the period 1850 - 2015. It should be noted that thermometers used to measure land or sea surface temperature were graduated in 1, 3, or 5°F increments until the mid-20th century. In addition, the report stated that the land surface air temperature has risen at twice the rate of the global average temperature due to the climate change. [13] It seems that the UNIPCC has migrated away from global warming to the term "climate change." Although global average temperature is a meaningless term in science, the term "climate change" is highly subjective and not subject to measurement, *per se.*

The "Special Report on Climate Change and Land" assigns the blame for every adverse change in the condition of the world's land mass to human-induced climate change, including the intensity of droughts, heavy precipitation events, severity of weather events, frequency and intensity of dust storms, desertification, vegetation greening, heat waves, animal pests, food insecurity and low crop yields, just to name a few. While the report encourages a change in a multitude of land use management practices to mitigate some of the effects of climate change, the socio-economic initiatives include empowering women, indigenous peoples, and other minorities in land use decision-making. The concluding statement encourages "rapid reduction in anthropogenic emissions of GHG across all sectors." [13] Like UNIPCC SR 15, the SRCCL offer no objective data to support any statements about the various aspects of the degradation of the condition of the Earth's land mass. Neither of these reports are scholarly scientific investigations of the subject matter. They are propaganda pieces designed to promote the global warming/climate change agenda.

It is obvious from each of the IPCC's statements above that the existence of AGW was considered a *fait accompli* from the inception of the organization in 1988. There was no effort to promote research to validate or falsify the hypothesis that man-made CO_2 emissions into the atmosphere had caused global warming. The only "scientific" studies were efforts to develop complex computer models, using coupled partial differential equations, to model the complex heat transfer interactions that occur among the Earth's various thermodynamic systems in the biosphere in an effort to predict future temperature changes related to changes in the CO_2 concentration in the atmosphere. Therefore, it is important to understand the construction of the mathematical models used in these computer programs because that understanding allows one to make a judgment on the efficacy of the model and its ability to predict climate outcomes.

There are a total of 65 climate models that are used by UN IPCC scientists that attempt to address the thermodynamic interactions involving energy and mass transfers with the Earth's land mass, atmosphere, and oceans.[11] The models not only cover multiple countries and continents, but there is often more than one model for a country. Scientists write coupled partial differential equations that often involve more than a million lines of complex computer code for a climate model to describe complex climate interactions. The results of the individual models are integrated into a "Unified Model" that attempts to predict global climate change using the results of the various models. Multiple Cray XC40 super computers capable of 14,000 trillion calculations per second are required to perform the computing. Sound complicated? The following is a *brief* explanation of the construct of the models and how they operate.

General Circulation Models (GCMs) utilize discrete equations for fluid motion and energy transfer and calculate the changes in the variables in these equations over time. GCMs divide the atmosphere and/or oceans into grids of discrete three dimensional "cells," which represent computational units. Unlike simpler models which make general assumptions about the net heat transfer or flux within a cell (related to thermodynamic interactions such as radiant heat flux or convection heat flux), those processes internal to a cell - such as convection heat transfer -that occur on scales too small to be resolved directly are *now calculated at the cell level by using a simplified process called parameterization, which specifies an average or* ***expected*** *effect of such processes on the resolved variables* .

The net heat flux is the rate of heat flow (usually expressed in $W \cdot m^{-2}$) that results from the interactions of the heat transfer mechanisms involved in the computational unit, such as radiant and convection heat transfer in the case of the atmosphere. Finally, the thermodynamic interactions between cells are calculated using "other functions." This is an important matter which is not explained in the IPCC literature. It determines the net heating or cooling that takes place in the thermodynamic system, and ultimately determines the calculation of the Earth's predicted average temperature.

It is important to understand that the objective of these models is to compute the net heat flux in the computational unit, in this case, a three-dimensional cell, and then sum these various heat flux values throughout the system to determine the overall change in heat flux, and ultimately temperature, within the system. To be able to compute this value with any reasonable degree of accuracy is highly unlikely.

Atmospheric GCM models set the thermodynamic boundary between the atmosphere and the ocean at the sea surface temperature. The models construct a series of interrelated partial differential equations that couple numerous factors that affect thermodynamic processes in both the ocean and the atmosphere including greenhouse gases like water vapor, CO_2, SO_2, CH_4, etc., along with heat transfer processes in the ocean. As noted above, *"(the models)* ***represent the pinnacle of complexity in climate models and internalize as many processes as possible*** *(emphasis added) and incorporate as many thermodynamic interactions as possible."* Simply stated, this sentence means that the IPCC equations represent a "macro approach" and contain as many terms as they can attempt to quantify (estimate). However, they state that the models are "still under development and *their degree of accuracy remains uncertain."* That would be an understatement!

Of particular note, in addition to developing these models that contain assumptions about many variables in the ocean and atmosphere, they also attempt to integrate into the models such variables as the "carbon cycle," which refers to the natural process by which the atmosphere and the world's oceans absorb and emit CO_2, as well as CO_2 exchanges in plant photosynthesis. There is very little, if any, scientific information that offers objective measurements of the numerous variables involved in the operation of the carbon cycle or photosynthesis. An effort to attempt to quantify any of those processes and integrate them into a predictive computer model would render the results meaningless at the outset.

Finally, they also contain assumptions about how changes in certain climate parameters over time would effect changes in other climate parameters over time (termed "positive and negative feedback") and the impact on temperature. These models are impossibly complex; they employ "the mother of all partial differential equations." And you know what? The UN IPCC brags about the fact! Apparently, none of the scientists who work at the IPCC have read Einstein's statement about the elegance of a simple equation. ***It would be an understatement to state that any effort to accurately estimate all of these inputs, their effect on other variables and their rates of change in such a dynamic environment as the Earth's biosphere would be an impossible task. The inclusion of these unproven assumptions renders the results meaningless at the outset.***

A key aspect of the IPCC models (it actually drives calculated temperature change and remains controversial) is what is referred to in the models as "climate forcings" and the concept of "positive and negative feedback." A climate forcing is an event, either natural or man-made, that affects the world's climate. The Earth's orbital cycle and volcanic eruptions are examples of natural forcings. Man-made emissions of CO_2 and CH_4 into the atmosphere are proposed by the proponents of AGW to be an example of anthropogenic climate forcings. [14]

Climate feedbacks are processes that result from climate forcings and can either amplify or diminish the effects of climate forcings. A feedback that increases an initial warming is called a "positive feedback." A feedback that reduces an initial warming is a "negative feedback."[14] For example, the modelers assume that a change (increase) in a climate parameter such as the concentration of CO_2 in the atmosphere causes an increase in heat energy absorbed and emitted back to Earth by the additional CO_2 molecules. This theoretical change in heat energy results in a radiative imbalance at the top of the Earth's atmosphere: the Earth's atmosphere retains the additional heat and global warming results. This event is termed a climate feedback by climate scientists.

However, the process does not end there.

A mathematical relationship among changes in the CO_2 concentration, heat retained in the atmosphere and water vapor concentration is then established in the model. For example, it is assumed that an increase in the concentration of CO_2 in the atmosphere causes an increase in the temperature of the troposphere. The warmer troposphere causes an increase in the concentration of water vapor in the troposphere, which causes a further increase in the temperature of the troposphere due to the increased absorption of LWIR emitted from the Earth's surface by the increased water vapor concentration. This results in an additional increase in water vapor concentration and the process repeats itself. This iterative process is known as a "feedback loop." *It should be noted that this is a theoretical concept which has not been proven (and would be difficult, if not impossible, to prove).*

It is proposed that a positive feedback loop causes a cycle of increasing temperature. A negative feedback loop causes a cyclical decrease in temperature. The accelerated formation of clouds because of an increase in the temperature and water vapor content of the troposphere is an example of a prospective negative feedback, since increasing cloud volume results in more of the sun's radiation being reflected back to the upper atmosphere. However, clouds (which contain water molecules) also absorb some of the LWIR emitted by the Earth's surface back to the atmosphere and retard the cooling of the surface. So, does increased cloud formation cause a positive or negative feedback? No one knows for certain. It depends on several different factors which would be difficult, if not impossible, to quantify.

The fundamental problem with the IPCC mathematical climate

models is that they contain hypotheses that have not been proven, such as the assumption that an increase in CO_2 concentration in the atmosphere causes an increase in water vapor concentration which triggers the positive and negative feedback mechanisms.

Further, the models are based on assumptions about energy and mass transfers within the Earth's biosphere that have not been proven, if it were even possible to do so. Finally, as we shall see in subsequent chapters dealing with energy transfer estimates involving coupled thermodynamic interactions, the range of uncertainty of the mathematical calculations often exceeds the values that result from the calculations, essentially rendering the results meaningless.

It is a commonly accepted practice in predictive mathematical modeling to conduct what is often called "back testing" on a newly developed model to validate the ability of the model to predict future unknown values. Back testing is the process whereby historical, known values for the variables contained in the model, are inserted into the model and the resultant calculated values (predictions) are compared to the actual, known historical values. Said another way, the model is tested to see if it can predict actual historical results using the actual values. If a predictive model fails the back testing process, it is normally discarded or modified in an effort to improve its predictive accuracy. **However, it seems that the UNIPCC does not view the ability of a model to pass a back test to be a prerequisite for use in climate predictions.**

Here is what the UN IPCC has to say about the effort to back test their climate models: [14]

"Although crucial *(emphasis added), the evaluation of climate models based on past climate observations has some important limitations.* ***By ne-***

cessity, it is limited to those variables and phenomena for which observations exist *(emphasis added). In many cases,* ***the lack or insufficient quality of long-term observations, be it a specific variable, an important process, or a particular region (e.g., polar areas, the upper troposphere/lower stratosphere (UTLS), and the deep ocean), remains an impediment*** *(emphasis added). In addition,* ***owing to observational uncertainties and the presence of internal variability, the observational record against which models are assessed is 'imperfect'*** *(emphasis added). These limitations can be reduced, but not entirely eliminated, through the use of multiple independent observations of the same variable as well as the use of model ensembles."*

"The approach to model evaluation taken in the chapter ***reflects the need for climate models to represent the observed behavior of past climate as a necessary condition to be considered a viable tool for future projections*** *(emphasis added). This does not, however, provide an answer to the much more difficult question of determining how well a model must agree with observations before projections made with it can be deemed reliable. Since the AR4, there are a few examples of emergent constraints where observations are used to constrain multi-model ensemble projections. These examples, which are discussed further in Section 9.8.3,* ***remain part of an area of active and as yet inconclusive research"*** *(emphasis added).*

In other words, what the IPCC stated above in Chapter 9 of the 2013 IPCC Report entitled, "Evaluation of Climate Models," is the following: we cannot conduct valid back tests on climate models because the historical data on many climate variables either does not exist or is not accurate.[11] As a result, we continue to make the effort to improve the ability of the models to predict future events; however, doubts about the ability of the models to accurately predict the future remain.

In conclusion, by their own admission the IPCC climate models cannot be validated; and, perhaps more importantly, an historical database for many important climate values is either incomplete or inaccurate. If that is the case, why should predictions of very small changes in the Earth's temperature be accepted by so many scientific organizations and academies as fact? **The truth is that proponents of the AGW hypothesis cannot base their conclusions on scientific data because none exists.**

In its 1996 report, the IPCC stated the following [15]:
"...decided to develop a new set of emissions scenarios which will provide input to the IPCC Third Assessment Report but can be of broader use than the IS92 scenarios. The new scenarios provide also (sic) input for evaluating climatic and environmental consequences of future greenhouse gas emissions and for assessing alternative mitigation and adaptation strategies. They include improved emission baselines and latest information on economic restructuring throughout the world, examine different rates and trends in technological change and expand the range of different economic-development pathways, ***including narrowing of the income gap between developed and developing countries*** *(emphasis added). To achieve this, a new approach was adopted to take into account a wide range of scientific perspectives, and interactions between regions and sectors."* As stated earlier, this was the first public mention of an IPCC agenda that included incorporating socio-economic criteria in climate models.

However, the IPCC statement above looks rather modest in its stated objective compared to the following from the U.N. Department of Economic and Social Affairs (UN/DESA). The UN/DESA describes itself as follows on its website: ***"UN DESA is a vital interface between global policies in the economic, social and environmental spheres***

and national action *(emphasis added). Its work is guided by the universal, integrated and transformative 2030 Agenda for Sustainable Development, along with a set of 17 Sustainable Development Goals and 169 associated targets adopted by the United Nations General Assembly on 25 September 2015." [16]*

The Development Policy and Analysis Division (DPAD) is a subset of the DESA. The Development Strategy and Policy Analysis Unit (DSP), which is a subset of the DPAD, states the following as its purpose: *"undertakes research on long-term issues relevant to the economic, social and environmental dimensions of development. The purpose of its research is to inform the development debates taking place around the UN Development Agenda and to provide advice to policy makers and development practitioners." [16]*

"UN Policy Brief #45: The Nexus between Climate Change and Inequalities," which was authored by the UN/DESA sub-unit of the DSP, states the following in the introduction to its report dated September 3, 2016 [16]: *"There is need for a better understanding on (sic) why climate hazards affect people unevenly owing to structural inequalities. This requires shifting from a narrow focus on identifying only the physical impacts of climate change, towards a broader analysis which also* ***incorporates the socioeconomic impacts of climate hazards*** *(emphasis added)."*

"The chapter examines the links between climate change and inequalities. More specifically, it shows that they are locked in a vicious cycle, whereby initial socioeconomic inequalities determine the disproportionate adverse effects arising from climate hazards, ***which in turn results in greater inequality*** *(emphasis added). This discussion is followed by a thorough review of the evidence demonstrating that the multiple dimensions of inequality*

(as they relate, inter alia, to ***income, assets, political power, gender, age, race, and ethnicity*** *(emphasis added)) underlie a situation where disadvantaged groups are more exposed and susceptible to climate hazards and possess less capacity to cope and recover when those hazards have materialized.* ***Part of the problem is that it took time for researchers across different disciplines to develop and then test the methodologies that allowed for a broadening of the focus to include socioeconomic impacts and the need to address them*** *(emphasis added)." [16]*

It seems that the members of the UN/DESA committee have taken a page out of the climate modeler's playbook. They have defined a feedback loop for socio-economic inequality: initial socio-economic inequalities determine the disproportionate adverse effects arising from climate hazards, which, in turn, results in greater inequality.

"The reality of the climate change-inequality vicious cycle has significant policy implications. ***The international community is committed to mitigating both climate change and inequality*** *(emphasis added)." [16]*

Upon its inception in 1988, the mission of the UN IPCC was primarily to be the *"clear scientific view on* ***the current state of knowledge*** *in climate change and its potential environmental and socio-economic impacts (emphasis added)·"* **As of September 2016, the stated mission is to address climate change within the context of eliminating the inequality among nations that exists along "multiple dimensions," such as income, assets, political power, gender, age, race, and ethnicity.** [16]

Instead of assessing the potential environmental and socio-economic impacts of climate change, **the mission of the UN IPCC and its sub-sets is to develop plans for action to address climate change**

that have as their objectives to eliminate inequality. Perhaps the comment that Christine Stewart, the former Minister of the Environment of Canada made in 1999 was prescient: *"No matter if the science is all phony, there are collateral environmental benefits... climate change provides the greatest chance to bring about justice and equality in the world." [17]*

UN Climate Treaties

On 1 April 2016, the United States and China, which together reportedly represent almost 40% of global emissions, issued a joint statement confirming that both countries would sign the Paris Climate Agreement. The Paris Climate Accord was "open for signature" effective April 22, 2016. It stipulated that *"this Agreement shall enter into force on the thirtieth day after the date on which at least 55 Parties to the Convention accounting in total for at least an estimated 55 per cent of the total global greenhouse gas emissions have deposited their instruments of ratification, acceptance, approval or accession."* [18] One hundred seventy-five Parties (174 states and the European Union) signed the agreement on the first date it was open for signature. [19] On August 24, 2016, President Obama signed the agreement on behalf of the U.S. The agreement was not submitted to the U.S. Senate for ratification.

On June 1, 2017, President Trump announced that the U.S. was withdrawing from the Paris Climate Accord, citing that the commitments of the prior administration to cut energy production and consumption would cost millions of U.S. jobs: *"would cut production for the following sectors: paper down 12 percent; cement down 23 percent; iron and steel down 38 percent; coal — and I happen to love the coal miners — down 86 percent; natural gas down 31 percent. The cost to the economy at this time would be close to $3 trillion in lost GDP and 6.5* *million industrial jobs, while households would have $7,000 less income and, in many cases, much worse than that."* [20] In addition, President Trump withdrew $2 billion of the Obama administration's commitment of $3 billion to the Green Climate Fund (GCF). The GCF describes its mission on its website as follows: *"GCF is a unique global platform to respond to climate change by investing in low-emission and climate-resilient development. GCF was established by 194 governments to limit or reduce greenhouse gas (GHG) emissions in developing countries, and to help vulnerable societies adapt to the unavoidable impacts of climate change." [21]*

The IPCC's mechanism for pursing this agenda is non-binding international agreements known as "climate treaties." While the stated purpose of these agreements is to reduce greenhouse gas emissions, they have failed to do so. They have also failed in their ulterior agenda of wealth transfer to developing nations.

The first such international agreement was the Kyoto Protocol on climate change, ratified by 36 countries in 2005. Vice-President Al Gore was a major proponent of the Kyoto Protocol and signed it on behalf of the Clinton administration. President Clinton did not submit the agreement to the U.S. Senate for ratification because "there was not meaningful participation by key developing countries in addressing climate change." [22] He was right.

The stated goal of the Kyoto Protocol was to reduce world-wide CO_2 emissions by 5% from their 1990 levels. **Instead, world-wide emissions of CO_2 increased by 32% from 1990 to 2010.** The Kyoto Protocol fell apart in 2012. No industrialized nation elected to participate in the second phase that was scheduled to run from 2012 - 2020. [22]

While the Kyoto Protocol did not introduce any new scientific proposals to reduce CO_2 emissions into the atmosphere, it did introduce a novel economic idea: the trading of carbon credits. This practice is often referred to as "cap and trade." A cap is set for a country's emissions; if a country's emissions exceed the established cap, compliance can be achieved through the trading of carbon credits. The result has been to create one of the largest and fastest-growing commodity markets in the world. The market for trading carbon credits has provided the opportunity for the enrichment of its market participants as well as the avoidance of the reduction of carbon emissions. These market participants include global investment firms supportive of progressive politicians and their climate-change agendas, and in some cases, the politicians themselves.

In 2005, the carbon market had a trading volume of about $11 billion. In 2019, world-wide carbon credits trading reached a record volume of $215 billion. Projections are that this market will continue to grow rapidly, with some estimating that trading volume could one day reach $1 trillion. The carbon market has created a new, and very lucrative segment in financial services. The largest carbon market participants are global investment firms. It should be no surprise that some of the biggest contributors to the campaigns of progressive politicians supporting a climate-change agenda are Wall Street investment houses who trade carbon credits. Some of those politicians have themselves jumped on this lucrative trading bandwagon. Generation Investment Management, a firm co-founded by former Vice-President Al Gore in 2004, is one of the largest private market participants. The fact that Mr. Gore is an owner and the chairman of a firm benefitting financially from an agreement he brokered on climate change would seem to be a potential conflict of interest.

Paris Climate Accord

The Paris Climate Accord (PCA) was designed by the IPCC to replace the failed Kyoto Protocol, with the same primary goals of emissions reduction, wealth transfer and renewing the Kyoto Protocol mechanism of carbon credit trading. Although the IPCC claims the PCA is legally binding, achieving emission limits is voluntary. There are no penalties for noncompliance with voluntary emissions targets, and no independent monitoring of the parties for accuracy in emissions reporting.

Why would any government sign such an agreement, when the effects of voluntary compliance - by sharply reducing the use of fossil fuel - would be devastating for employment, the economy, energy costs, and the reliability of the nation-wide power grid? A related question is why did the IPCC promote the Paris Climate Accord when the Kyoto Protocol demonstrated that emissions cap and trade agreements are not effective in reducing CO_2 emissions? The answer is that trading carbon credits is big business, and a tremendous wealth-generator for backers of the climate-change agenda. **If one wants to understand the purpose behind the AGW hypothesis, follow the money.**

So, just what exactly is the Paris Climate Accord (PCA)? Let's review its basic tenets: [18]

1. It states as an objective to address socio-economic inequalities in the context of addressing the threat of climate change.
2. It states that the Parties should not do anything to threaten world food production.
3. Sets a goal to keep the rise in the Earth's mean temperature less than 2°C and preferably 1.5°C.
4. Requires each Party to communicate a "nationally determined contribution target" to lower carbon emissions every five years.

5. The Parties will consider the economic impact of climate change mitigation efforts on developing countries.
6. The Parties will preserve CO_2 sinks (trees and other vegetation) and pay countries not to cut trees.
7. Establishes the concept of "internationally transferred mitigation outcomes) that would be "voluntary and authorized by participating Parties." A participating Party (Country 1) could either buy carbon emission allowances (carbon credits) from Country 2; or take actions to lower emissions in Country 2. Then Country 2 could transfer the resulting carbon credits to Country 1 to offset carbon emissions in Country 1.
8. The "Conference of the Parties," who are the signatories to the Agreement, would get a share of the value of the "internationally transferred mitigation outcomes to cover administration costs and assist developing countries vulnerable to the adverse impacts of climate change." In addition, the Conference of the Parties agreed to adopt rules as to how the money from "internationally transferred mitigation outcomes" would be shared.
9. The "developed countries" agreed to "mobilize" $100 billion by 2020 to start-up the PCA activities and contribute $100 billion per year until 2025.

What are the Parties that are signatories to the PCA required to do as part of the agreement?

1. Mandatory, legally binding emission reductions by the Parties? No. The goals are voluntary.
2. Comply with an enforcement mechanism to ensure achievement of the "nationally determined contribution target" to lower carbon emissions every five years? No. There is no enforcement mechanism.

3. Pay economic penalties associated with non-compliance? No. There are no economic penalties for non-compliance.
4. Engage in an audit process to insure accurate reporting of carbon emissions? No. There is none.
5. Engage in an audit process to ensure that the funds allocated to the Conference of Participating Parties from "internationally transferred mitigation outcomes" are properly expended? No. There is no accountability system in place.

In conclusion, the PCA is a very loosely worded document whose primary aim is to accomplish the aforementioned goals of the UN IPCC: transfer wealth from the developed nations to the undeveloped nations under the guise of addressing climate change and eliminate the inequality among nations that exists along "multiple dimensions," such as income, assets, political power, gender, age, race, and ethnicity.

CHAPTER SIX

THE SCIENTIFIC METHOD OF INQUIRY

WHEN IT COMES to the "science" of the AGW hypothesis, it appears that adherence to the principles of the "Scientific Method" have been abandoned by many practitioners in the field. In this chapter, we will review the principles and operation of the scientific method; and develop a complex hypothesis within the construct of the scientific method to evaluate the claims of AGW.

Historically, scientific research has adhered to the scientific method of inquiry. The scientific method is defined as: *"An empirical method of acquiring knowledge that has characterized the development of science since at least the 17th century. It involves careful observation and applying rigorous skepticism about what is observed, given that cognitive assumptions can distort how one interprets the observation. It involves formulating hypotheses, via induction, based on such observations; and experimental and measurement-based testing of deductions drawn from the hypotheses. The final step of the scientific method requires refinement (or elimination) of the hypotheses based on the experimental findings."* [1]

The scientific method requires that its practitioners follow a discipline for conducting research:

1. Careful observation of a scientific phenomenon applying rigorous skepticism about what is observed, given that cognitive assump-

tions can distort how one interprets the observation. Said another way, one must test the legitimacy of the observation to ensure that it is an accurate description of the phenomena observed and is not biased due to cognitive assumptions, i.e., the observer presupposes a desired outcome.

2. Construction of a hypothesis or set of hypotheses that clearly and accurately state a supposition or proposed explanation made on the basis of limited evidence as a starting point for further investigation.
3. Develop a framework for an experiment(s) or other investigations to verify or falsify the hypothesis or set of hypotheses.
4. Measure the results of the experiment(s). **Implicit in the validity of the scientific method as a means to conduct research is the ability of the researcher (and others) to objectively confirm or refute (falsify) the hypothesis by replicating results of the experiment.**
5. Refine or eliminate the hypothesis. The results of the experiment or investigation may cause one to refine the hypothesis to include (or eliminate) factors that do or do not appear to have a causative effect on the phenomena observed. Of course, **the hypothesis should be eliminated (discarded) if the results of the experiment disprove (falsify) the hypothesis.**

William Thomson (June 26, 1824 – December 17, 1907), First Baron Kelvin, often referred to simply as Lord Kelvin, was an Irish mathematical physicist who famously stated, *"I often say that when you can measure what you are speaking about, and express it in numbers, you know something about it; but when you cannot measure it, when you cannot express it in numbers, your knowledge is of a meagre and unsatisfactory kind; it may be the beginning of knowledge, but you have scarcely, in your thoughts, advanced to the stage of science, whatever the matter may be."* [2]

An important aspect of a scientific hypothesis is that it be falsifiable. I list below a quote from the presentation of a paper prepared by Francico Ayala entitled, *Darwin and the Scientific Method,* that was presented to the National Academy of Sciences of the U.S. in January 2009 [3]:

"The requirement that scientific hypotheses be falsifiable rather than simply verifiable seems surprising at first. It might seem that the goal of science is to establish the "truth" of hypotheses rather than attempt to falsify them, but it is not so. There is an asymmetry between the falsifiability and the verifiability of universal statements that derives from the logical nature of such statements. A universal statement can be shown to be false if it is found to be inconsistent with even one singular statement, (i.e., a statement about a particular event). ***But a universal statement can never be proven true by virtue of the truth of particular statements, no matter how numerous these may be*** *(emphasis added)."*

I find this aspect of the scientific method to be quite fascinating. It is the marriage of science and philosophy and serves as the basic underpinning of the integrity of the scientific method. *To qualify as a legitimate scientific hypothesis, there must be a means to attempt to prove the hypothesis wrong.* Obviously, if over time, rigorous testing of the hypothesis does not contradict its assertions, then it gains credibility and may rise to the level of a scientific theory.

The discipline and objectivity that has historically characterized research in the fields of science such as physics and thermodynamics are embedded in the methodological rigor of the scientific method of inquiry. It is that discipline, objectivity, and adherence to methodological rigor that has produced the many breakthroughs in science over the years. Basic scientific research, especially in the work of scientists like

Sir Isaac Newton, James Maxwell, and Albert Einstein, was an inquiring mind's effort to advance man's knowledge of the physical world. There seemed to be no political or social agenda in the pursuit of the research. As an aside, none of the aforementioned received funding from any governmental body or organization for their research. Both Newton and Maxwell came from families of comfortable means. While Einstein held several academic posts during his lifetime, he did not require a research budget per se. His mind was his laboratory and his blackboard the medium to express his thinking!

The interesting thing about all this is that a *hypothesis is always subject to the possibility of being falsified at a point in time.* A case in point is Einstein's General Theory of Relativity. Developed in 1915, the *General Theory of Relativity* postulated the geometric theory of gravitation. In Einstein's view, gravity was a geometric property of space and time; or spacetime. Einstein's view was that the universe is characterized by a curvature of spacetime which is directly related to the energy and momentum of whatever matter and radiation are present. [4]

Einstein's *General Theory* predicted several phenomena which took years to confirm. One, gravitational lensing, is the "bending" of light as the result of the gravitational forces of matter (such as a cluster of galaxies, known as a "gravitational lens") between a distant light source and an observer. Based on the *General Theory,* Einstein predicted that the amount that light would be deflected by the gravitational force of the Sun was twice the value predicted by Newtonian mechanics. This hypothesis was proven correct by Eddington and Dyson using an astrographic lens during a total solar eclipse in 1919. [5] It was not until 1979 that the first gravitational lens would be discovered. When asked by his assistant what his reaction would have been if *General Relativity* had not been confirmed by Eddington and Dyson in 1919, Einstein said, *"Then I would feel sorry for the dear Lord. The theory is correct anyway."* [5]

As I have engaged in my research into the subject of the AGW hypothesis, one aspect of that research has been abundantly clear to me: **the proponents of the AGW hypothesis have accepted the validity of the hypothesis without sufficient research or data to prove it; or, perhaps more importantly, they have accepted the validity of the hypothesis without any attempt to falsify it.** It seems that any effort to question the hypothesis is viewed as the equivalent of scientific heresy. Unlike Einstein's hypotheses, which were and still are subjected to rigorous, objective testing over decades, the AGW hypothesis has, in some quarters, become sacrosanct, a *fait accompli.*

The purpose of this rather detailed academic analysis of the scientific method is to demonstrate that research that pertains to the subject of man-made global warming should be subjected to scientific rigor (i.e., efforts to falsify the hypotheses). In my opinion, that is the test for the presence of the three-legged stool of legitimacy in scientific research: methodological rigor, exactitude and objectivity based on the scientific method. If the data (or conclusions) have not been derived employing these principles, then the efficacy of the scientific study must be questioned.

If one were to follow the logic above, it would be expected that a climate scientist who believed that man had caused global warming that would adversely affect life on Earth in the future, would state the following hypothesis: *"Man has caused global warming that will result in future climate conditions that will adversely affect life on Earth."* However-er, the hypothesis above actually contains three conjectures which must be developed into a complex hypothesis:

1. "Global warming has occurred; that is, the temperature of the world's oceans, land mass and relevant atmosphere has risen during the period under investigation by a statistically significant amount."
2. "Man's activities are responsible for the global warming that has occurred."
3. "The extent to which global warming has occurred, or is reasonably projected to occur in the future, will adversely affect life on Earth."

If any of the conjectures in the complex hypothesis above are found to be invalid, the complex hypothesis is rendered null.

In the following chapters of this book, we will review the scientific research and the data that is available to falsify or validate each of the conjectures in the complex hypothesis above.

CHAPTER SEVEN

THE LAWS OF THERMODYNAMICS

GIVEN THE NATURE of the claims of the proponents of the man-made global warming hypothesis, the science of thermodynamics should be central in governing any investigations of the subject matter. In essence, thermodynamics is the science that deals with the movement of heat, which is central to the matter of global warming. If investigations into the subject of anthropogenic global warming are to have any scientific credence, they must adhere to the precepts of the "first principles of science" concept. The laws of thermodynamics and the mechanisms of heat transfer establish those first principles and any analysis or conclusions about global warming must agree with them [1]

Thermal energy is the most fundamental form of energy in our solar system. All living things on Earth depend on thermal energy for life. Based on our current understanding of thermodynamics, the universe itself could not function without heat. Mass, down to the smallest particle in the universe, cannot have motion without heat energy. It characterizes the nature of stars, planets, and other celestial bodies. Life would not be possible on Earth without our heat source - the Sun. In conjunction with the Earth's atmosphere, the Sun dictates the temperature of the Earth's surface, oceans, and the lower atmosphere (troposphere) where life exists. The Earth's climate is the result of the solar energy that Earth receives and the various thermodynamic processes that occur within the Earth's biosphere: its land mass, oceans, and lower troposphere. Therefore, in order to understand how the Sun's solar energy affects our climate, we need to review the laws of thermodynamics and the mechanisms of heat transfer to gain a basic understanding of the scientific principles involved. Such an understanding will enable us to employ scientific laws, deductive reasoning, and common sense to understand the impact that each thermodynamic interaction might have on the Earth's climate.

Thermodynamics is the branch of physics that deals with the relationship between heat and other forms of energy (such as mechanical, electrical, or chemical energy), and by extension, the relationships among all forms of energy. [1,2] You may know that there are four laws of thermodynamics; what you might not know is the sequence in which they were developed. Around 1850, Rudolf Clausius (German scientist and a central founder of thermodynamics) and William Thomson (First Baron Kelvin, Scottish mathematical physicist, and engineer) stated both the First Law and the Second Law of Thermodynamics. [3,4] The Third Law was developed by the German chemist Walther Nernst during the years 1906 to 1912. [5] The Zeroth Law was stated by Sir Ralph H. Fowler (a British physicist and astronomer) who invented the title *"The Zeroth Law of Thermodynamics"* in 1939, when he was discussing the 1935 work of two Indian scientists, entitled, "A Treatise on Heat." [6,7] Since it was considered so elemental to the physics of thermodynamics, scientists decided that it needed to be inserted before the First Law. So, let's examine each of the Laws of Thermodynamics to understand what they mean and how they apply to heat transfer on our planet.

The Zeroth Law of Thermodynamics states that, *"if two thermodynamic systems are each in thermal equilibrium with a third, then they are in thermal equilibrium with each other. Two systems are said to be in the relation of thermal equilibrium if they are linked by a wall (boundary) permeable only to heat and they do not change over time."* [6] Although it is not stated in the Zeroth Law per se, I would add what I consider to be an important corollary to this Law; that is, if two adjacent systems are not in thermal equilibrium and they share a boundary permeable to heat, thermal energy spontaneously transfers from the system with higher temperature to the system with lower temperature, *ceteris paribus.*

While this corollary seems elemental, it forms the basis for the fundamental science that underpins the concept of heat transfer. If an open system like the Earth (where matter and energy can cross the system boundary) is not in thermal equilibrium (the temperature is not constant within the entire system), heat will be transferred within the system from the warmer body to the cooler body. In Earth's case, if the temperature of the land mass is greater than the temperature of the ocean, then heat will flow from the land mass to the ocean across a shared thermal boundary until they reach thermal equilibrium. If the temperature of the land mass is greater than the troposphere, then heat will flow from the land mass to the troposphere across a shared thermal boundary until thermal equilibrium occurs. If the temperature of the troposphere is greater than the stratosphere, then heat will flow from the troposphere to the stratosphere until they are in thermal equilibrium.

In the Mesosphere, a portion of the atmosphere at an altitude of around 60 miles, the temperature is -80°C. Therefore, a thermal gradient (a decrease in temperature as altitude increases) should be present in the atmosphere where heat moves from the hotter air mass to the cooler air mass and into space. As a practical matter, in an open thermodynamic system, thermal equilibrium does not exist except on a local basis. Constant exchanges of mass and energy within the system cause continuous changes in the temperature of the various thermodynamic components in the system. The path and rate of heat transfer is a function of the temperature differential and the area and thermal permeability of shared boundaries. Let's use a thought experiment to illustrate this concept.

Suppose we have a friend who has a house in Manteo, North Carolina, on the beach. Let's take a trip there and go for a swim in the ocean in September. The mean ocean temperature off Manteo in September is reported to be 25.6°C or, 78.1°F. We will assume that our body has a temperature of 98.6°F. That is a temperature differential of 20.5°F, so when we enter the ocean, heat will be transferred from our body to the ocean (don't worry, the sea level probably won't rise). If we just stick our toe into the water, the area of the toe (boundary area) that is in contact with the water controls the rate of heat transfer – it is relatively small, and we don't feel much heat loss. However, if we were to wade into the water up to our waist and increase the area of the shared boundary (more skin exposed to the cooler water), the rate of heat transfer would increase as a function of the increase in the boundary area, *ceteris paribus.* If we stayed in long enough and the temperature of the ocean remained below our body temperature, our body would continue to give up heat until it reached thermal equilibrium with the ocean. Warning: Do not test this Law, it is harmful to your health!

The First Law of Thermodynamics (Law of the Conservation of Energy): Energy cannot be created or destroyed in an isolated system. [3] An isolated system is a thermodynamic system that cannot exchange either energy or matter outside the boundaries of the system.

Most theoretical physicists today believe that the universe is expanding. Therefore, our universe may have the characteristic of being a "finite infinity." That is, the universe may have a boundary that is constantly expanding. The fact of the matter is that we do not know if the universe is infinite or not. The observable universe has an estimated radius of 46 billion light years! Given that we believe the universe to be expanding, the observable universe is classified as an open thermodynamic system. That is, matter and energy can and do cross the boundary of the observable universe. These definitions will take on more meaning for us as we attempt to analyze the multitude of heat transfer actions that take place within the Earth's biosphere on a constant basis. The fact that both energy and matter can enter or exit the Earth's thermodynamic systems is critical to our understanding of the issue of the "science" of AGW.

The Second Law of Thermodynamics: I have seen the Second Law written in several different phrasings. I list three of those below: [4]

1. The Second Law of Thermodynamics states that the state of the entropy of the entire universe, as an isolated system, will always increase over time. The Second Law also states that the changes in the entropy in the universe can never be negative.
2. The Second Law of Thermodynamics also states that in any cyclic process, entropy will either increase or remain the same (if all processes are reversible). Entropy is also a measure of the amount of thermal energy in a system which is unavailable to do work.
3. The Second Law of Thermodynamics states that in all energy exchanges, if no energy enters or leaves the system, the potential energy of the state will always be less than that of the initial state.

Finally, an important aspect of the Second Law is known as the "Clausius statement of the Second Law," which was developed by Rudolph

Clausius in 1854: *"No process is possible* ***whose sole result*** *(emphasis added) is the transfer of heat from a cooler to a hotter body."* [4] In his statement, Clausius makes a provision for the concept of work in a thermodynamic system to cause heat transfer from a colder to a warmer body, such as occurs in refrigeration (in 1834, the first working vapor-compression refrigeration system was built). [8]

An understanding of the path of the flow of heat is fundamental to the conceptual understanding of thermodynamics. **In the absence of work, heat always flows from a warmer body to a cooler body.** As we will review in later chapters when discussing heat transfer in the atmosphere, scientific analyses by climate scientists in the examination of the thermodynamic processes involved in the atmosphere sometimes fail to consider this important aspect of the Second Law.

It seems to me that the concept of entropy has almost a metaphysical quality to it. In essence, the concept of entropy in a thermodynamic system states that there is no absolute conversion of one form of energy to another. When heat energy is converted to kinetic energy, some amount of heat is lost to the system and is not available to do work. It is a thermodynamic quantity representing the unavailability of a system's thermal energy for conversion into mechanical work, described as entropy, or as the *degree of disorder or randomness in the system.*

A common example of an increase in the entropy of a system is the thermodynamic process involved in an internal combustion engine in an automobile. When the fuel/air mixture is ignited in the combustion chamber (cylinder volume), the burning of the fuel creates a hot gas (air with products of combustion) that expands to drive the piston connecting rod, causing the crankshaft to turn, resulting in a force transmitted through the power train that ultimately drives the wheels of the car. However, immediately upon ignition, some of the heat in the combustion chamber is lost through the cylinder walls, engine block and exhaust gases to the environment. In addition, heat loss to the environment occurs at every point where there is friction in the drive train as the mechanical forces work to turn the car's wheels. Each of these events creates heat that is lost to the environment and increases the entropy of the universe.

On a macro scale, it is said that the entropy of the universe is expanding. Theoretical physicists employ this concept to establish a "thermodynamic arrow" which supports the idea of a cooling universe over time as it continues to expand. Let's consider this concept in some more detail and have a little fun. Since thoughts have no mass, we can perform a thought experiment where we travel to the edge of the universe at many times the speed of light (it is a long way there, wherever "there" is). When we get there, we "peer over the edge" to the point where at 0 K, all thermodynamic activity has ceased. The most advanced cryogenic radiometer instruments available today indicate that the temperature of cosmic background radiation in the universe is 2.725 K. The temperature of the cosmic background radiation is expected to continue to decrease as the universe expands. 0 K is absolute zero; it can't get any colder!

In concert with the Third Law of Thermodynamics, entropy is constant at 0 K. [5] The universe is no longer expanding. We can see light, in the form of discrete particles or waves, depending on your view of quantum mechanics, perhaps "frozen" in space. Everything appears at a standstill. However, as far as we know, one law remains unaffected by all this. The Universal Law of Gravity. I wish Sir Isaac Newton was here with us. All matter has mass and the forces of gravity still apply even at 0 K. Slowly, imperceptibly, the densest matter in the universe, like black holes and neutron stars, exert gravitational force to attract matter near them. They absorb that matter, increasing their mass and the force of their gravitational field as they combine. The process accelerates. Entropy takes on a negative value in violation of the Second Law. The universe begins to collapse into itself - The Big Crunch - followed by the Big Bang? Welcome to the concept of gravitational singularity, as it relates to the fate and rebirth of the universe. Some theoretical physicists think this may happen 15 billion to 30 billion years from now. Who knows?

The Third Law of Thermodynamics states that the entropy of a system approaches a constant value as the temperature approaches absolute zero (0 K).[5] We can't seem to get away from the concept of entropy. It is also stated in the following way: *"Because a temperature of absolute zero is physically unattainable, the entropy of a perfect crystal approaches zero as its temperature approaches absolute zero. We can extrapolate from experimental data that the entropy of a perfect crystal reaches zero at absolute zero, but we can never demonstrate this empirically." [5]*

Finally, I offer to you some quotes on the importance of thermodynamics from some imminent physicists of the past [9]:
"A theory is the more impressive the greater the simplicity of its premises, the more different kinds of things it relates, and the more extended its area of applicability. Therefore, the deep impression that classical thermodynamics made upon me. It is the only physical theory of universal content which I am convinced will never be overthrown, within the framework of applicability of its basic concepts." - Albert Einstein, Autobiographical Notes (c.1940s)

"The future belongs to those who can manipulate entropy; those who understand but energy will be only accountants." - Frederic Keffer

"Every mathematician knows it is impossible to understand any elementary course in thermodynamics." - Vladimir Arnold, "Contact Geometry: The Geometrical Method of Gibbs' Thermodynamics" (1989)

Heat Transfer Mechanisms

There are three main heat transfer mechanisms: conduction, convection, and radiant. We can easily demonstrate these three mechanisms with a thought experiment. [10]

Assume that it is a sunny day, and the ambient temperature is around 85°F, and you decide to walk outside in your bare feet. To get a better view, you walk onto the asphalt driveway in front of your house and away from the trees. The sun is shining, the sky is clear and there is a gentle breeze. What heat transfer mechanisms do you experience?

Conduction Heat Transfer

Assuming that the asphalt is not so hot that you can't continue to stand on it, you experience conduction heat transfer from the asphalt into your bare feet. Conduction occurs when two bodies at different temperatures (T of asphalt > 98.6°F) are in contact with each other. Heat flows from the warmer to the cooler object until they are both at the same temperature (thermal equilibrium). Conduction is the movement of heat through a substance by the collision of molecules (a molecule is composed of two or more atoms; two hydrogen atoms and one oxygen atom combine to form a molecule of water). At the place where the two objects touch, the faster-moving molecules of the warmer object (the heat gives them more kinetic energy) collide with the slower moving molecules of the cooler object. As they collide, the faster molecules give up (transfer) some of their energy to the slower molecules. The slower molecules gain heat and more kinetic energy; then, they collide with other molecules in the cooler object. This process continues until heat energy from the warmer object spreads throughout the cooler object, the molecules have the same thermal energy level and thermal equilibrium is achieved. [10]

Convection Heat Transfer

While you are standing on the asphalt (or even if you step off it), a gentle breeze begins to blow. You experience the sensation of cooling on your exposed skin. This is convection heat transfer. In liquids and gasses, convection is usually the most efficient way to transfer heat. In this case, convection is the process of losing heat through the movement of cooler air molecules across the skin - the cooler air molecules receive a transfer of heat from the warmer molecules at the skin surface (same molecular energy exchange mechanism as described above). Convection also occurs when warmer areas of a liquid or gas rise to cooler areas in the liquid or gas due to differences in their respective densities. As this happens, cooler liquid or gas takes the place of the warmer areas which have risen higher. This cycle results in a continuous circulation pattern and heat is transferred to cooler areas. You see convection when you boil water in a pan. The bubbles of water that rise are the hotter parts of the water rising to the cooler area of water at the top of the pan. [10]

Radiant Heat Transfer

While you are standing on the asphalt (or otherwise), you feel the sun shining on you and its warmth. This is what is known as radiant heat transfer. Both conduction and convection require matter to transfer heat. Radiation is a method of heat transfer that does not rely upon any contact between the heat source and the heated object. For example, we feel heat from the sun even though we are not touching it. Heat can be transmitted through space by thermal radiation. Thermal radiation is a type of electromagnetic radiation. Radiation is a form of energy transport consisting of electromagnetic waves traveling at the speed of light. No mass is exchanged, and no medium is required. [10]

Electromagnetic radiation is a stream of particles, called photons, traveling at the speed of light in the form of a wave. Each electromagnetic wave contains a certain amount of energy dependent upon its wavelength (we will elaborate on the properties of waves below). Radiant heat transfer is the result of the energy transfer of radiation waves of differing wavelengths and therefore differing energy potential. [10]

It is time for another thought experiment. Think about the trip we took to Manteo, North Carolina, when we went swimming in the ocean to learn about what affects the rate of heat transfer. Suppose that while we were knee deep, an occasional wave passed by us towards the shore with a certain wave height (amplitude). In this case, the amplitude represents the amount of kinetic energy (force) the wave possessed. Since we obviously have some superpowers, we can see the peak of the second wave that is coming towards us at exactly the moment the peak of the first wave strikes us, and we very accurately calculate the distance from peak to peak. That measurement is known as the "wavelength." Suddenly, the wind velocity increases. The waves are still at the same height (amplitude), with the same kinetic energy (force) but now they are coming in faster, striking us more often (shorter wavelength). The shorter the wavelength between wave "strikes," the more energy we absorb. It is getting harder to remain upright. As a result of our thought experiment and the resulting experience, we deduce that waves of a shorter wavelength (with the same amplitude such as photons) have more energy. Congratulations! You now know the fundamental science behind the concept that different waves in the electromagnetic spectrum possess different energy levels. Why is this important?

The electromagnetic waves that the sun emits have different wavelengths and therefore, different energy levels (heat energy). Some of those waves don't make it all the way to Earth's surface. They may get reflected by clouds or absorbed by water vapor in the Earth's atmosphere. In addition, once they reach the Earth's surface, they may either be reflected by the Earth's surface or absorbed by it. Why is this important? It has to do with the theoretical concept of the "Earth's Energy Budget," and it impacts what climate scientists call the "greenhouse effect." We will study this issue in more depth in a later chapter.

In conclusion, electromagnetic energy/photon energy is solely a function of the photon's wavelength. Unlike our experience with the waves in the ocean, the "amplitude" of a photon does not change. When dealing with "particles" such as photons or electrons that have small amounts of energy, a commonly used unit of energy is the electron-volt (eV) rather than the joule (J). An electron volt is the energy required to raise an electron through one volt; a photon with an energy of $1 \text{ eV} = 1.602 \times 10^{-19}$ J.

There are three other heat transfer mechanisms that occur in the atmosphere that we will consider in our analysis of the thermodynamic interactions within the Earth's thermodynamic systems. Latent (often referred to as "hidden") and sensible heat are types of energy released or absorbed in the atmosphere. Latent heat is related to changes in phases between liquids, gases, and solids. Sensible heat is related to changes in temperature of a gas or object with no change in phase.

Latent Heat

Latent heat is energy released or absorbed, by a body or a thermodynamic system, during a constant-temperature process, usually a first-order phase transition like a change from a liquid to a vapor. Latent heat can be understood as energy which is supplied or extracted to change the state of a substance without changing its temperature. Examples are *latent heat of fusion* and *latent heat of vaporization* involved in phase changes, i.e., a substance condensing or vaporizing at a specified temperature and pressure. In *latent heat of vaporization,* heat is transferred from the environment to the substance in a process known as an endothermic reaction. In *latent heat of condensation,* heat is transferred from the substance to the environment, in a process known as an exothermic reaction. In either case, the temperature of the substance does not change, only that of the environment. In meteorology, *latent heat flux* is the flow of energy (usually expressed as Watts per meter squared or $W \cdot m^{-2}$) from the Earth's surface to the atmosphere that is associated with evaporation or transpiration of water at the surface, and subsequent condensation of water vapor in the troposphere.[11]

Sensible Heat

Sensible heat is heat exchanged within a thermodynamic system that changes the temperature of the system without changing some variables such as volume or pressure. As the name implies, sensible heat is the heat that you can feel. The sensible heat possessed by an object is evidenced by its temperature.[12]

Differential Equations and Predictive Climate Models

In our review of the mathematical models used by climate scientists to try to predict the Earth's temperature change that results from changes in other climate parameters, we will discuss the employment of differential equations. Differential equations are often used by scientists to describe or predict natural phenomena.

A differential equation is defined as "a mathematical equation that relates some function with its derivatives. In applications, the functions usually represent physical quantities, the derivatives represent their rates of change, and the equation defines a relationship between the two. Because such relations are extremely common, differential equations play a prominent role in many disciplines including engineering, physics, economics, and biology." [13]

Climate scientists use partial differential equations in climate models to define a relationship (known as a function) between a change in a dependent variable, such as temperature, and the change in one or more independent variables, such as the CO_2 concentration in the Earth's atmosphere. A derivative of a variable represents the rate of change of that variable with respect to the rate of change of another variable (often time). Ordinary differential equations contain derivatives with respect to only one variable, partial differential equations contain derivatives with respect to more than one variable. Let's review an example that illustrates the nature of derivatives to better understand how these equations operate. We will use the example of a person who waits on a traffic light to change to cross the street.

In physics, the mechanical definition of an object that is stationary is that the object has position; in this case, a person who is standing still on a street corner waiting to cross the street. When the traffic light changes, the person begins to walk at a constant rate, known as velocity. The first derivative of position is velocity, which represents the change in position with respect to the change in time. While the person is crossing the street, the light begins to flash yellow, warning that it will soon change to green. The person increases the velocity at which they walk. The second derivative of position is acceleration; acceleration is the first derivative of velocity.

In climate science, differential equations ***assume*** a relationship between a dependent variable and two or more independent variables and then calculate the effect of a change in one or more independent variables on the dependent variable. The following is an ordinary differential equation that relates the change in the Earth's temperature over time (dependent variable) to a change in the radiant heat flux in the atmosphere (independent variable) times a constant:

$\partial T/\partial t=(\Delta Rf)\ \lambda$, where T= temperature, t=time, $\partial T/\partial t$ = change in temperature over time, ΔRf = change in radiant heat flux and λ = climate sensitivity factor constant. This equation ***assumes*** the following: first, that the rate of change in temperature over time is related to the product of the change in heat flux and the climate sensitivity factor constant; and second, that the climate sensitivity factor constant accurately represents the amplifying effects of the change in heat flux on the change in temperature.

Suppose that we believed that other factors might affect the change in the Earth's temperature over time, such as the water vapor concentration in the atmosphere, the gross domestic product, and the population of the country to which the climate model applies? We would add a term for each of those independent variables to create a partial differential equation as follows:

$\partial T/\partial t=(\Delta Rf)\ \lambda+(\partial Wv/\partial t)\ k+(\partial GDP/\partial t)\ k+(\partial P/\partial t)\ k$, where Wv = water vapor concentration, k = a constant, GDP = gross domestic product and P = population. In each case, we would have to estimate a value for the constant and the independent variables: radiant heat flux, the water vapor concentration, the gross domestic product, and the population of the country as well as **estimate** their rates of change.

The extent to which the estimates for one or more of the values of the variables or their rates of change are incorrect will determine the accuracy of the resultant overall calculation, which is the sum of the terms of the equation. Now imagine how much more difficult it would be to construct a partial differential equation (more than likely coupled partial differential equations) that would accurately describe changes in the Earth's temperature as a function of a multitude of thermodynamic interactions involving the land mass, atmosphere, and the world's oceans, based on population, GDP, land use, gender mix, political power, energy cost, and other socio-economic factors. How could one possibly identify and populate the terms with the correct values, given the dynamic nature of all the variables involved?

To avoid the complexity stated above, a weakness in scientific modeling often involves simplifying the equations used to a reduced set that can be solved for specific conditions. In climate science, an example would be the use of the equilibrium climate assumption. Here, the time dependent solar flux is replaced by a long-term 24-hour average value. However, this assumption does not represent the reality of what occurs in nature and can produce CO_2 induced warming as a mathematical artifact in the climate model. Most climate modeling has been based on the *a-priori* assumption that an increase in atmospheric CO_2 concentration must produce global warming.

Scientific Notation

We will be dealing with very large numbers that are involved in certain values associated with the Earth's thermodynamic systems; and dealing with very small numbers related to certain thermodynamic interactions within those systems. Therefore, it will be helpful to review the basics of scientific notation.

Scientific notation is the way that scientists easily write and manipulate very large or very small numbers. For example, instead of writing 0.0000000056, we write 5.6×10^{-9}. On the other hand, 5,600,000,000 would be expressed as 5.6×10^{9}. In the case above, when the sign of the exponent is negative, the decimal point is moved to the left nine places, adding eight zeros to denote the full number. If the sign of the exponent is positive, then the full value of the number would be expressed by moving the decimal point to the right eight places, adding eight zeros. It is merely a way to display the value of a number without zeros; and it facilitates shorthand calculations as noted below.

To add two numbers expressed in scientific notation, first, the exponents must be equal. For example:

$$(56 \times 10^{3}) + (3.5 \times 10^{4}) = (5.6 \times 10^{4}) + (3.5 \times 10^{4}) = 9.1 \times 10^{4}$$

Subtraction requires equivalent exponents as well. For example:

$$(56 \times 10^{3}) - (3.5 \times 10^{4}) = (5.6 \times 10^{4}) - (3.5 \times 10^{4}) = 2.1 \times 10^{4}$$

When multiplying numbers expressed in scientific notation, one must multiply the base numbers, then add the exponents as follows:

$$(56 \times 10^{3})\,(3.5 \times 10^{4}) = (56 \times 3.5) \times 10^{(3+4)} = 196 \times 10^{7}$$

...which can be expressed as either 19.6×10^8 or 1.96×10^9. Usually, the least number of whole numbers to the left of the decimal point, the easier to calculate.

Dividing numbers expressed in scientific notation requires that the base numbers be divided, and the exponents be subtracted as follows:

$$(56 \times 10^6) / (3.5 \times 10^4) = (56 / 3.5) \times 10^{(6-4)} = 16 \times 10^2, \text{ or } 1.6 \times 10^3$$

We will make extensive use of scientific notation in subsequent chapters involving heat transfer calculations and quantum mechanical analysis.

CHAPTER EIGHT

THE THERMODYNAMIC INTERACTIONS WITH THE EARTH'S LAND MASS

THE CONCEPT of the Earth's average temperature is an abstraction and has no meaning in the physical world or in scientific analysis. It is a figment of the climate scientist's imagination, conjured up to prove a flawed hypothesis. A calculation of equal value would be to determine the average zip code in the United States (49663 - Manton, MI, population 1,555) to locate the average American city. Nevertheless, for years, climate scientists have endeavored to determine the average temperature of the Earth through various methodologies.

Before we review the methodology employed to calculate the average surface temperature of the Earth, I think that a review of basic mathematical concepts is in order to understand the flaw in the logic employed. Mathematics is the language of science; however, *it is important that scientists who employ mathematics not seek to mislead or deceive.* A calculation in mathematics can be accurate without being valid. Let me illustrate this point with a simple example.

Suppose that during the 2020 - 2021 coronavirus pandemic, a principal in a school asked a teacher to report on the health of the fifteen students in his classroom. The teacher decided to measure the body temperature of each student as a proxy for their health condition. Seven of the students had a body temperature of 97.6°F; five had a temperature of 97.1°F and three had a temperature of 103°F, because they had recently contracted the coronavirus. The average temperature of the fifteen students was calculated to be 98.5°F, well within the average range for a "normal" body temperature. Accordingly, the teacher reported to the principal that the health of the students was normal. The mathematical calculation was accurate, but the conclusion was invalid. Let's examine the methodology that climate scientists use to calculate the average land surface temperature to illustrate the illegitimacy of the methodology employed to calculate the Earth's average temperature.

There are four different world organizations that attempt to measure global mean temperature. The GISS is an acronym for the Goddard Institute for Space Studies. GISS is located at Columbia University in New York City and maintains a land surface temperature database. The institute is a laboratory in the Earth Sciences Division of NASA's Goddard Space Flight Center and is affiliated with the Columbia Earth Institute and School of Engineering and Applied Science. [1] The Hadley Centre is part of the British Meteorological Office (Met Office) research center. The Met Office Hadley Centre for Climate Change, "named in honour of George Hadley, is one of the United Kingdom's leading centres for the study of scientific issues associated with climate change. It is part of, and based at, the headquarters of the Met Office in Exeter." [2] The most recent iteration of the land surface temperature database is referred to as "HadCRUT4." The UAH temperature dataset for the troposphere is maintained at the University of Alabama-Huntsville, as part of an ongoing joint project between UAH, NOAA, and NASA.[3] Remote Sensing Systems (RSS) in Santa Rosa, California, is a company supported by NASA for the analysis of satellite data used to develop a temperature database of the troposphere. [4] It is unclear why NASA funds both the UAH and the RSS efforts to develop a temperature dataset for the troposphere.

The GISS and Hadley Centre both gather ground-based temperature readings from meteorological surface air temperature stations (MSATS). The data from the MSATS is obtained by measuring the air temperature in an enclosure placed 1.5 m – 2 m above the ground. The MSATS readings depend on the temperature of the bulk air mass of the local weather system, surface LWIR heat flux, air convection and wind speed. [5] In addition to the aforementioned variables, the advent of "urban heat islands" – an urban or metropolitan area that is significantly warmer than its surrounding rural areas due to human activities - can cause a warming bias in the temperature readings. There tends to be a weak correlation between the MSATS and measured actual ground temperature.

Average surface air temperature anomalies for the month and year are calculated at a given MSATS station location based on the following procedure: 1) record the minimum and maximum temperature for each day 2) calculate the daily average temperature by adding the minimum and maximum temperature and dividing by two 3) calculate the average for the month from the daily data; 4) calculate the annual average by averaging the monthly data; 5) the average monthly or annual temperature is compared to a 30–year history of each and the variance from the historical average is calculated and expressed as a "temperature anomaly" for the month or year. An average monthly or yearly temperature anomaly that is higher than the historical average is known as a positive anomaly, indicating warming; if it is lower, it is a negative anomaly, indicating cooling. [5]

The fundamental error in this approach is the averaging of the temperature readings. Temperature in thermodynamic terms is defined as the measure of the average kinetic energy of the molecules in the system (mass with a defined boundary). It relates specifically to a property of the system at the moment in time that the temperature is observed and is related to the ability of the system to transfer heat energy to an adjacent system. Therefore, the daily average temperature is a meaningless term in scientific analysis; it represents a mathematical calculation that has no counterpart in the real world. Ergo, the same reasoning applies to calculating the yearly average temperature; it is a meaningless term. An even more egregious error would be to average the temperatures of two different locales, since heat transfer occurs across adjacent thermal boundaries, not between two distant locales.

So, what happens when the data for a location between reporting stations is not available, the raw data does not conform to a range of historical readings, or the temperature readings are taken at a time of day that is different from prior readings? Scientists analyzing the raw data interpolate and "adjust" the data.

The raw data recorded from the temperature stations is not simply used in the averaging calculations; first, it is adjusted. The National Climate Data Center (NCDC) is the agency primarily charged with adjustments to the raw data in the U.S., although Berkeley Earth also adjusts and analyzes the raw data, as well. Different agencies use different adjustment methods. The station data is adjusted for homogeneity (i.e., nearby stations are compared and adjusted if trends are different). [6,7]

The U.S. based NCDC data set is freely available, including both raw and adjusted data, so that anyone can see what the adjusted and unadjusted station data shows. **However, neither the raw data nor the adjusted data produced by HadCRUT4 and used by the IPCC was publicly available until after the "Climategate" scandal in November 2009.** [8]

Previously, HadCRUT4 had only published a list of stations and the calculated 5 x 5-degree grid anomaly results. However, the homogeneity adjustments can have a significant effect on the temperature readings and trends, as can be seen in Figure 1 below. One can see that the difference between the Tmax and Tmin adjustments can be as much as 0.6°C. In addition, it should be noted from the graph that since 1980, the homogenization adjustments have resulted in a warming bias from 0.1 to 0.4°C. [7]

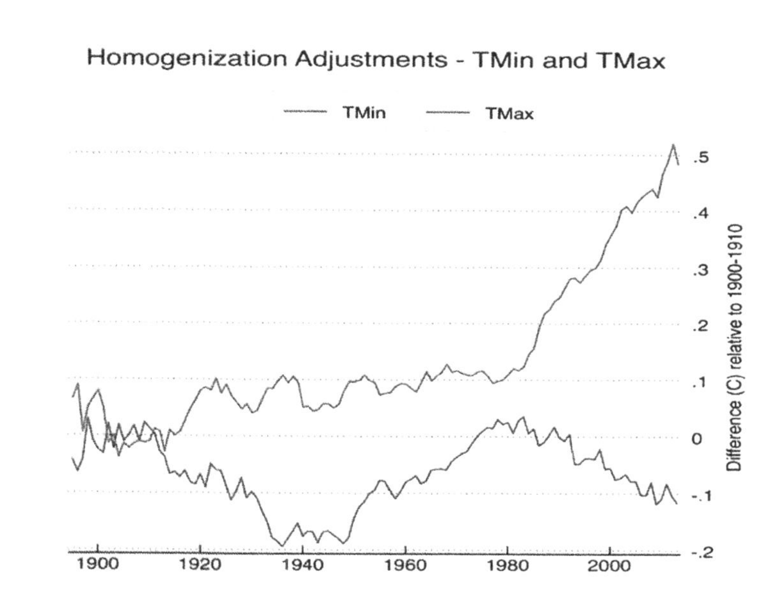

Figure 1. Pairwise Homogenization Algorithm adjustments to USHCN relative to the 1900 to 1910 period. Courtesy of NCDC. [7]

On balance, it is difficult to determine what the individual effects of homogeneity adjustments have on the temperature datasets due to the inherent variability of the method of the adjustment of the raw data over time. The adjustments can include a provision for location (site) changes, changes in the equipment used to obtain the temperature readings at a site, "micro-site" changes such as the movement of the measuring device, "macro-site" changes such as the relocation of the measuring device or the development of a "urban heat island" (large urban area) close to the measuring station.

Another significant action that is taken to adjust the data is what is known as the "Time of Observation" adjustment, which refers to what time of day the temperature reading was obtained. Temperature observations are gathered by a volunteer network in the U.S. As seen in Figure 2, the graph illustrates the adjustments made to the minimum and maximum temperature readings due to the time of day of the observations. [5] It should be noted that beginning around 1980, the adjustments for the time of observations have steadily risen from a positive adjustment range for Tmax of 0.02°C to 0.2°C to a positive adjustment range for Tmin of 0.05°C to 0.27°C. For example, in the year 2000, the adjustments to the maximum and minimum temperatures actually recorded added 0.15°C to the maximum daily temperature and 0.20°C to the daily minimum temperature, creating a warming bias to the average yearly temperature record. It is clear that Time of Observation adjustments can have a material warming biasing effect on the average temperature readings, the value of which can be significantly greater than the measured temperature variance.

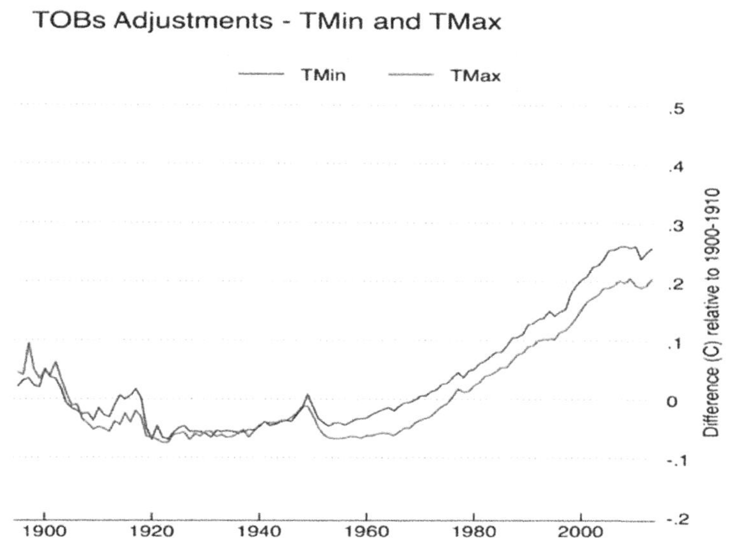

Figure 2. Impact of TOBs adjustments on U.S. minimum and maximum temperatures via USHCN. Courtesy of NCDC. [3]

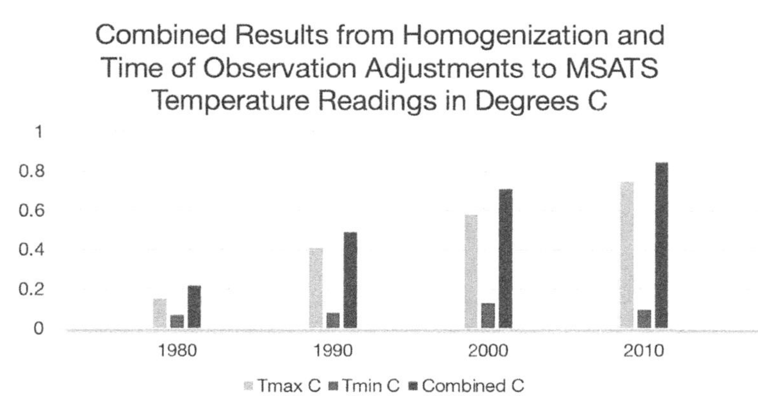

Figure 3. The Combined Results from Homogenization and Time of Observation Adjustments to MSATS Maximum and Minimum Temperature Readings in °C.

Figure 3 above depicts the combined results from adjustments to Homogenization and Time of Day Observations to the MSATS temperature readings for period 1980 – 2010. In its first report in 1990, the UN IPCC stated: *"Under the IPCC Business-as-Usual (Scenario A) emissions of greenhouse gases, a rate of increase of global mean temperature during the next century of about 0.3°C per decade (with an uncertainty range of 0.2°C to 0.5°C per decade), is greater than that seen over the past 10,000 years..."* [9]

It should be noted that the sum of the homogeneity and time of day adjustments during the period 1980 to 2010 have increased in value in each succeeding decade. These adjustments have contributed an increase in average MSATS temperature readings of 0.56°C per decade for the period, well in excess of the UN IPCC predictions of 0.3°C. In effect, adjustments to temperature readings have artificially "created" global warming.

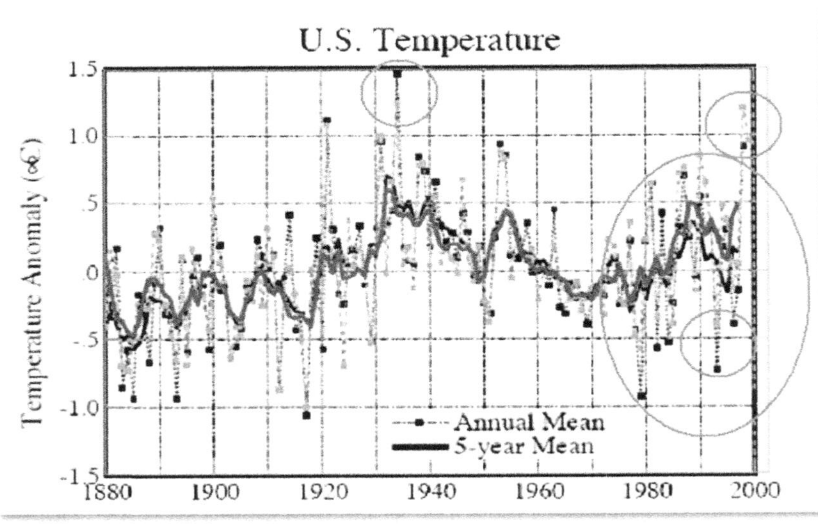

Figure 4. Comparison of U.S. Temperature Changes Due to Change in Adjustment Methods [10] Courtesy of GISS NASA.

Figure 4 demonstrates how an increase in temperature trend was achieved simply by changing the method of adjusting the data. Some of the major changes are highlighted in this figure – such as the decreases in the 1930s and the increases in the 1980s and 1990s. A comparison of U.S. temperature datasets and temperature trends from Hansen et al are shown. The resulting adjusted temperature dataset from HadCRUT4 U.S. stations (red) and adjusted temperature dataset from GISS updated U.S. temperatures (blue) are shown. On this graphic, base period of datasets were set equal for the period 1930 to 1939. [10]

In its 2014 study, the IPCC concluded, "Each of the last three decades has been successively warmer at the Earth's surface than any preceding decade since 1850. In the Northern Hemisphere, 1983 to 2012 was likely the warmest 30-year period of the last 1,400 years (emphasis added)." [11] However, as can be seen from Figure 5, **the unadjusted temperature anomaly for the period 1920-1950 was much higher than the period 1980-2000, in direct contradiction to the 2014 IPCC statement.**

Figure 5 shows a more recent example of the HadCRUT (Hadley Centre/Climate Research Unit Temperature) re-adjustment of Global Average Surface Temperature data. This adjustment was made for historical data reported for the period 1850-2010. The graph clearly depicts adjustments to the temperature readings in the 1930's-1950's which eliminates the warming trend first measured in the unadjusted data (see Figure 4).[10] In addition, the cooling trend depicted in the unadjusted data in Figure 4 for the period 1950-1980, was eliminated as shown in Figure 5, by the addition of an artificial warming trend created through data adjustments of 0.10 - 0.17°C.[12] Perhaps most striking, is the added warming from 1950 -1970 to offset what was originally measured to be a cooling period, as depicted in the unadjusted data in Figure 4. Also, warming was added for the period 1900-1930 to remove prior reported cyclicality. Finally, from 1980 onward, warming has been added to reinforce the appearance of a sharp increase in global warming. **The obvious conclusion is that both NASA/GISS and HadCRUT have manipulated daily temperature readings and adjusted historical temperature data to remove cycles in the temperature data, eliminate cooling periods and create the appearance of global warming.**

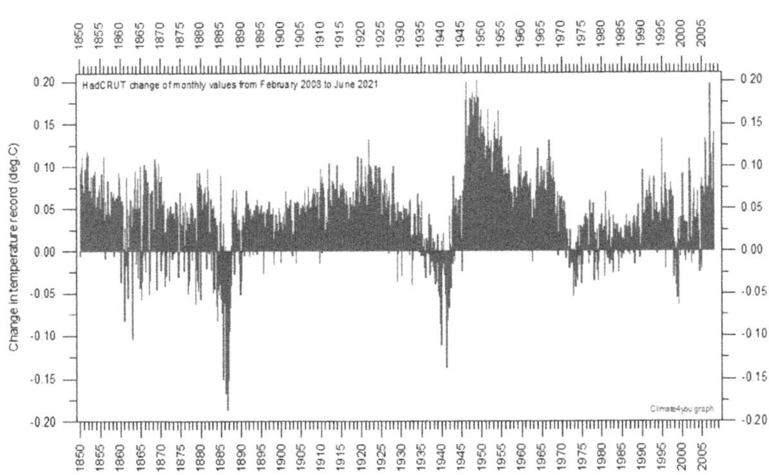

Figure 5. HadCRUT Change in Historical Monthly Surface Air Temperature Values for the Period 1850-2010. Courtesy of NCAR UCAR. [12]

NASA has consistently adjusted its data and/or interpolated data into grid areas that have no monitoring stations. Those actions have biased the purported accuracy of any global mean temperature calculations. The amount of the increase is roughly equivalent to the increase in global mean temperature reported by the IPCC for the period.

The Audit of the HadCRUT4 Temperature Database

In June 2018, Dr. John McLean, PhD., published a report entitled, *"An Audit of the Creation and Content of the HadCRUT4 Temperature Dataset."* The publication was based on his doctoral thesis (awarded in December 2017 by James Cook University, Townsville, Australia) which focused on an audit of the HadCRUT4 database and associated files as of January 2016. [13]

"The report uses January 2018 data and revises and extends the analysis performed in the original thesis, sometimes omitting minor issues, sometimes splitting major issues and sometimes analyzing new areas and reporting on those findings. The thesis was examined by experts external to the University, revised in accordance with their comments and then accepted by the University. This process was at least equivalent to "peer review" as conducted by scientific journals." [13]

Dr. McLean summarizes his findings in the "Executive Summary" of the report: *"As far as can be ascertained, this is the first audit of the HadCRUT4 dataset, the main temperature dataset used in climate assessment reports from the Intergovernmental Panel on Climate Change (IPCC). Governments and the United Nations Framework Convention on Climate Change (UNFCCC) rely heavily on the IPCC reports so ultimately the temperature data needs to be accurate and reliable."* [13]

This audit shows that it is neither of those things (emphasis added).

More than 70 issues are identified, covering the entire process from the measurement of temperatures to the dataset's creation, to data derived from it (such as averages) and to its eventual publication. The findings (shown in consolidated form Appendix 6) even include simple issues of obviously erroneous data, glossed-over sparsity of data, significant but questionable assumptions and temperature data that has been incorrectly adjusted in a way that exaggerates warming. [13]

It finds, for example, an observation station reporting average monthly temperatures above 80°C, two instances of a station in the Caribbean reporting December average temperatures of 0°C and a Romanian station reporting a September average temperature of -45°C when the typical average in that month is 10°C. On top of that, some ships that measured sea temperatures reported their locations as more than 80km inland. It appears that the suppliers of the land and sea temperature data failed to check for basic errors and the people who create the HadCRUT dataset did not find them and raise questions either. [13]

The processing that creates the dataset does remove some errors, but it uses a threshold set from two values calculated from part of the data, but errors were not removed from that part before the two values were calculated. Data sparsity is a real problem. The dataset starts in 1850 but for just over two years at the start of the record the only land-based data for the entire Southern Hemisphere came from a single observation station in Indonesia. At the end of five years just three stations reported data in that hemisphere. Global averages are calculated from the averages for each of the two hemispheres, so these few stations have a large influence on what's supposedly "global." [13]

Related to the amount of data is the percentage of the world (or hemisphere) that the data covers. According to the method of calculating coverage for the dataset, 50% global coverage was not reached until 1906 and 50% of the Southern Hemisphere was not reached until about 1950. [13]

In May 1861, global coverage was a mere 12% - that's less than one-eighth. In much of the 1860s and 1870s most of the supposedly global coverage was from Europe and its trade sea routes and ports, covering only about 13% of the Earth's surface. To calculate averages from this data and refer to them as "global averages" is stretching credulity.

Another important finding of this audit is that many temperatures have been incorrectly adjusted. The adjustment of data aims to create a temperature record that would have resulted if the current observation stations and equipment had always measured the local temperature. Adjustments are typically made when a station is relocated or its instruments or their housing replaced. The typical method of adjusting data is to alter all previous values by the same amount. Applying this to situations that changed gradually (such as a growing city increasingly distorting the true temperature) is very wrong and it leaves the earlier data adjusted by more than it should have been. Observation stations might be relocated multiple times and with all previous data adjusted each time the very earliest data might be far below its correct value and the complete data record show an exaggerated warming trend.

The overall conclusion (see Chapter 10) is that the data is not fit for global studies. Data prior to 1950 suffers from poor coverage and very likely multiple incorrect adjustments of station data. Data since that year has better coverage but still has the problem of data adjustments and a host of other issues mentioned in the audit.

Calculating the correct temperatures would require a huge amount of detailed data, time, and effort, which is beyond the scope of this audit and perhaps even impossible. **The primary conclusion of the audit is however that the dataset shows exaggerated warming and that global averages are far less certain than have been claimed** (emphasis added).

One implication of the audit is that climate models have been tuned to match incorrect data, which would render incorrect their predictions of future temperatures and estimates of the human influence of temperatures (emphasis added). Another implication is that the proposal that the Paris Climate Agreement adopt 1850 to 1899 averages as "indicative" of pre-industrial temperatures is fatally flawed. During that period global coverage is low – it averages 30% across that time – and many land-based temperatures are very likely to be excessively adjusted and therefore incorrect. A third implication is that even if the IPCC's claim that mankind has caused the majority of warming since 1950 is correct then the amount of such warming over what is almost 70 years could well be negligible. The question then arises as to whether the effort and cost of addressing it make any sense. **Ultimately it is the opinion of this author that the HadCRUT4 data, and any reports or claims based on it, do not form a credible basis for government policy on climate or for international agreements about supposed causes of climate change."** (emphasis added) [13]

The conclusions that McLean reaches in his report are damning to say the least. Interestingly, instead of trying to defend the methodology or accuracy of the data, the U.K. Meteorological Office and the Hadley Centre's response was that they welcomed the audit and *"any actual errors identified will be dealt with in the next major update."* [14]

The audit by McLean was not the first time that the climate scientists at the Climate Research Unit at the University of East Anglia, UK, had experienced a credibility problem. In November 2009, it was discovered that over 1,000 emails from the scientists at the Climatic Research Unit had been hacked. The hacked emails suggested a conspiracy by some climate scientists to withhold data that did not support the global warming hypothesis, restrict public access to data and adjust data to support the global warming hypothesis. [15]

The Earth's Energy Budget

An important component of the AGW hypothesis concerns the Earth's Energy Budget, which is a hypothetical model that depicts the flow of energy from sunlight as it enters the Earth's atmosphere. The AGW hypothesis is based in part on the concept that the absorption and downward re-emission of LWIR photons by an increased concentration of CO_2 in the Earth's atmosphere creates an energy imbalance. That energy imbalance is caused by the Earth's absorption of the downwelling LWIR re-emitted by the increased CO_2. that would have otherwise flowed back out of the atmosphere and into space. Therefore, less heat energy flows back out through the atmosphere than enters it. The difference between the heat flows is the amount of heat energy that the Earth purportedly absorbs which causes global warming. This aspect of the AGW hypothesis is based on the First Law of Thermodynamics which states that energy can neither be created or destroyed. There are several flaws in the thinking involved in the development of this aspect of the AGW hypothesis. To understand those flaws, it is necessary to understand the fundamental flaws involved in the construct of the Earth's Energy Budget.

Figure 6 is one of many similar diagrams that have been published by various science organizations to depict what happens to sunlight that enters the Earth's atmosphere.[16] We will return to an analysis of the Energy Budget above. However, I think that it will be helpful to our analysis to study the history of the development of the Earth's Energy Budget in more detail.

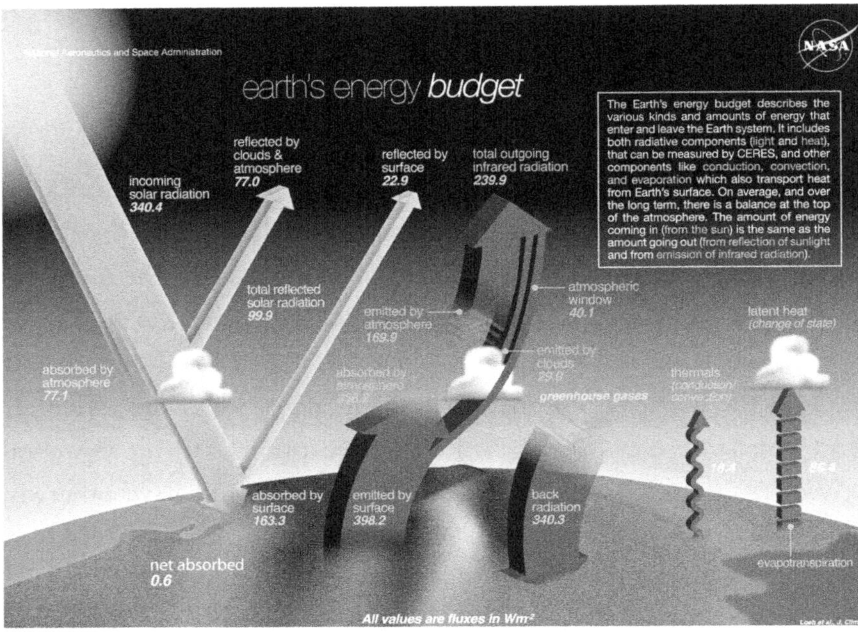

Figure 6. Earth's energy budget describes the balance between the radiant energy that reaches Earth from the sun and the energy that flows from Earth back out to space. Credits: NASA [16]

The Earth's Energy Budget

The following history of the development of the "Energy Budget" is from a 2013 publication of the *Astronomical Society of the Pacific* [16]: *"For Earth, scientists have been working since the early 1900s to lay out a diagram summarizing the key elements of this budget and to quantify its various components. Hunt et al (1986) summarizes work in this area prior to the satellite era. Beginning with the launch of Nimbus-7 in 1978, scientists began to study the energy budget from space. In particular, a series of NASA instruments have been dedicated to understanding and monitoring the energy budget since the mid-1980s: The Earth Radiation Budget Ex-*

periment (ERBE) and the Clouds and the Earth's Radiant Energy System (CERES). Kiehl and Trenberth (1997) published an energy budget diagram that has become something of a standard. It uses data from satellites, ground-based instruments, aircraft field campaigns, and ***computer models*** *(emphasis added) to estimate the magnitude of each type of energy flow."*

The aforementioned language would give one the impression that the Earth's energy budget has been developed over time with a great deal of scientific exactitude. While sincere efforts have undoubtedly been made to that end, the current version of the Earth's energy budget is rife with assumptions that renders its analysis virtually useless.

As is stated in the First Law of Thermodynamics, energy can neither be created nor destroyed. The concept of the Earth's energy budget is based on this law; that is, the solar energy in the form of radiated heat energy that enters the top of Earth's atmosphere must be equal to that which exits the top of the atmosphere. That fact is undoubtably true *over time in an open thermodynamic system (the Earth),* where mass and energy can cross its boundaries. However, this statement assumes that the Earth has an equilibrium temperature, which is not the case. The temperature of the Earth is different at every point in time and space. In addition, different thermodynamic systems such as the Earth's land mass, oceans and atmosphere have different abilities to store heat, known in technical terms as heat capacity. So, the land mass or world's oceans might absorb heat energy for a period of time and then transfer that energy back to the atmosphere and into space at a later time. Finally, what happens to the Sun's radiant heat energy as it moves through the atmosphere is unknown. I have listed below the NASA analysis of the Energy Budget diagram:

"Earth's energy budget describes the balance between the radiant energy that reaches Earth from the Sun and the energy that flows from Earth back out to space. Energy from the Sun is mostly in the visible portion of the electromagnetic spectrum. About 30 percent of the sun's incoming energy is reflected back to space by clouds, atmospheric molecules, tiny, suspended particles called aerosols, and the Earth's land, snow, and ice surfaces. The Earth system also emits thermal radiant energy to space mainly in the infrared part of the electromagnetic spectrum. The intensity of thermal emission from a surface depends upon its temperature." [17]

"The UCAR Center for Science Education (UCAR SciEd) serves the geoscience community by amplifying and complementing the work of the National Science Foundation's National Center for Atmospheric Research (NCAR) and University Corporation for Atmospheric Research (UCAR) member universities by reaching our audiences: K-12 educators, university faculty, students, and the public through excellence in educational programs and experiences." [18]

Here is what UCAR has to say about the factors that affect the Earth's energy budget:
"Many different things cover the Earth such as soil, rocks, water, forests, snow, and sand. Materials like these have different ways of dealing with the solar energy that gets to our planet. Dark colored surfaces, like ocean and forests, reflect very little of the solar energy that gets to them. Light colored parts of the planet surface, like snow and ice, reflect almost all of the solar energy that gets to them." [19]

"The amount of energy reflected by a surface is called albedo. Albedo is measured on a scale from zero to one (or sometimes as a percent). Very dark colors have an albedo close to zero or close to 0%. Very light colors have

an albedo close to one or close to 100%. Because much of the land surface and oceans are dark in color, they have a low albedo. They absorb a large amount of the solar energy that gets to them, reflecting only a small fraction of it. Forests have low albedo, near 0.15. Snow and ice, on the other hand, are very light in color. They have very high albedo, as high as 0.8 or 0.9, and reflect most of the solar energy that gets to them, absorbing very little." [20]

"The albedo of all these different surfaces combined is called the planetary albedo. ***Earth's planetary albedo is about 0.31*** *(emphasis added). That means that about a third of the solar energy that gets to Earth is reflected out to space and about two thirds is absorbed. The Moon's albedo is 0.07, meaning that only 7% of the energy that gets to it is reflected."* [20]

"If Earth's climate is colder and there is more snow and ice on the planet, more solar radiation is reflected back out to space and the climate gets even cooler. On the other hand, when warming causes snow and ice to melt, darker colored Earth surfaces and oceans are exposed, and less solar energy is reflected out to space causing even more warming. This is known as the ice-albedo feedback. Clouds have an important effect on albedo too. ***They have a high albedo*** *(emphasis added) and reflect a large amount of solar energy out to space. Different types of clouds reflect different amounts of solar energy. If there were no clouds, Earth's average albedo would drop by half."* [20]

If one were to search the internet for scholarly papers that purported to analyze the Earth's energy budget, the results of that research would demonstrate that there is a variance of opinion (and it is just that) as to what happens to the sunlight that enters the top of the atmosphere as it makes its way to Earth. While satellites can measure the amount of solar energy from the Sun that reaches the top of Earth's atmosphere with some exactitude (often called the solar constant ~1365 $W \cdot m^{-2}$), how that energy is dispersed as it enters the Earth's various atmospheric layers and in what quantities is pure speculation. While speculation may form part of a hypothesis, it has no place in scientific analysis.

The above analysis utilizes assumptions about average values of albedo (ability of a body to reflect sunlight), and the subsequent heat transfer values of the various interactions of the thermodynamic systems. Not only is it impossible to know these values with any reasonable degree of accuracy, but the requirement to use average values in the analysis when such a wide degree of variance in those values occurs, renders the analysis useless. As we shall see, the range of those values can be very broad, and any assumptions made about the average value can have a significant effect on the calculated value of the heat flux. **This weakness in this analysis is of particular importance when one attempts to compare the relatively small net heat flux from the addition of CO_2 to the Earth's atmosphere to the overall energy budget of the Earth.**

Interestingly, at another website address, NASA has this to say about calculating the Earth's energy budget: *"Determining exact values for energy flows in the Earth system is an area of ongoing climate research.* ***Different estimates exist, and all estimates have some uncertainty*** *(emphasis added). Estimates come from satellite observations, ground-based observations, and numerical* ***weather models*** *(emphasis added)."* [21]

As will be demonstrated at the conclusion of this analysis, no scientific investigations to date have produced measured data describing how energy flows as it moves through the atmosphere towards the Earth and then back out into space. An inherent difficulty in trying to develop such an analysis is that atmospheric conditions such as cloud cover and relative humidity are constantly changing, presenting enormous challenges to the development of meaningful mathematical models for such a dynamic system.

Most scientists use an approximate mean value of 1,365 $W{\cdot}m^{-2}$ for the solar irradiance that reaches the top of the Earth's atmosphere; this value has been derived from satellite measurements. To develop the Earth's energy budget, climate scientists have found it necessary to make assumptions about the *average* amount of sunlight that reaches the top of the Earth's atmosphere at any point in time. This average considers the fact that the Earth is a sphere, solar illumination varies in space and time, approximately half of the Earth is in darkness every day, incoming solar energy varies considerably from tropical latitudes to polar latitudes, and the angle of incidence of solar radiation is constantly changing because the Earth's axis of rotation is tilted 23.5 degrees relative to its plane of eccentric orbit around the Sun. [21]

To consider all of the aforementioned variables, climate scientists used the following reasoning to develop the *average* fraction of the solar irradiance that reaches the top of the Earth's atmosphere over time in order to develop the Earth's energy budget. The surface area of a sphere (the Earth's geometric shape) is four times the cross-sectional surface area of a circular disk. Only half of the Earth's surface receives solar illumination at any point in time. [21]

In addition, *"the total solar irradiance is the maximum power the Sun can deliver to a surface that is perpendicular to the path of incoming light. Because the Earth is a sphere, only areas near the equator at midday come close to being perpendicular to the path of incoming light. Everywhere else, the light comes in at an angle. The progressive decrease in the angle of solar illumination with increasing latitude reduces the average solar irradiance by an additional one-half. Therefore, the average energy from the Sun that reaches the top of the atmosphere at any point in time is one-quarter of the solar constant, or approximately 341* $W{\cdot}m^{-2}$*."* [21] Simply stated, about one-half of the Earth is in darkness at any point in time. Then, because the surface of the Earth is curved, no point in time (other than the Earth's equator, which is roughly perpendicular to the Sun's rays) receives the full value of the remaining one-half of the total value of radiation. Therefore, based on geometric calculations, the Earth receives about one-half of the remaining one-half of sunlight. Therefore, the average amount of solar energy reaching the top of the atmosphere is one-fourth the total value, or approximately 341 $W{\cdot}m^{-2}$.

Depending on the source, estimates vary as to what happens to the estimated 341 $W{\cdot}m^{-2}$ that, on average, reaches the top of the Earth's atmosphere. Most scientific studies state that around 29% (100 $W{\cdot}m^{-2}$) of the incoming solar radiation is reflected back into space, as represented in Figure 8. It should be noted that "clouds and the atmosphere" are represented as reflecting 77 $W{\cdot}m^{-2}$, or about 23%, and the Earth's surface is depicted as reflecting 23 $W{\cdot}m^{-2}$, or about 7%.

These calculations make significant assumptions dealing with albedo, the proportion of the incident light or radiation that is reflected by the surface of a body; and the magnitude of clouds present in the atmosphere. For example, it is estimated that cloud albedo can vary from 10-90%, depending on a number of factors such as liquid water or ice content, thickness of the cloud and the angle of incidence of sunlight." [22, 23] It should be noted that cloud formation within the Earth's atmosphere is a function of many variables and is constantly changing.

Therefore, it is impossible to quantify the average cloud content of the Earth's atmosphere over time or the average cloud albedo. Any estimate of the average amount of sunlight reflected by clouds is essentially a guess. Perhaps more importantly, like so many other efforts to analyze thermodynamic interactions within the Earth's biosphere, the use of an average value of a climate parameter like cloud albedo to calculate the Earth's average energy budget (a meaningless term in and of itself), cannot produce an accurate result. The range of prospective values is so broad so as to render the calculation invalid. It is by its very nature a contradiction of logic and is *prima facia* evidence of its falsity.

Regarding the albedo of the Earth's surface, the estimated range of values is from 10-40%. Snow has an estimated albedo of around 80-90%, sand around 40% and the ocean around 6%. [23] The only other variable of any significance in calculating the Earth's average surface albedo would be the amount of snow and ice present, which of course, is constantly in flux. However, since the value of the Earth's estimated surface albedo heat flux and reflected energy is relatively low (23 $W{\cdot}m^{-2}$ or about 7%), the estimated amount of snow and ice plays a smaller role in efforts to determine heat transfer interactions within the Earth's thermodynamic systems. Let's continue with our review of the NASA analysis of the Earth's Energy Budget [23]:

"Less than half of the incoming sunlight heats the ground (emphasis added). The rest is reflected away by bright white clouds or ice or gets absorbed by the atmosphere." **Again, it should be noted that there have been no direct measurements to determine any of these values, and they are in a constant state of flux.** These values are merely estimates that have been developed in an effort to conform to the requirement of the principle of the conservation of energy that is contained in the First Law of Thermodynamics. Since it is assumed that an average of 341 $W{\cdot}m^{-2}$ of energy enters the top of the Earth's atmosphere, then the same amount of energy, *ceteris paribus,* must exit the top of the Earth's atmosphere to maintain an energy balance. The assignment of certain energy values to the various thermodynamic interactions is an effort to comply with that dictate.

Perhaps most importantly, the concept of the heat flows involved in the construct of the Earth's Energy Budget ignores the heat capacity of the Earth's land mass and oceans. Heat capacity is defined as the amount of heat that is required to raise the temperature of the mass of a substance (like the Earth's land mass or oceans) by 1°C. Both the Earth's land mass and oceans can "store" a significant amount of heat for a period of time and later release it to the atmosphere and into space. Therefore, the Earth's land mass, oceans and atmosphere are never in thermal equilibrium and there is never a short-term balance between the heat entering the top of the Earth's atmosphere and that exiting it. Like the concept of an equilibrium temperature of the Earth's surface, the concept of a balanced heat flow that enters and exits the top of the Earth's atmosphere over the short term is a figment of the climate scientists' imagination.

In conclusion, the Earth's Energy Budget makes assumptions about the amount of cloud cover, the albedo of the cloud cover, the value of surface albedo and the heat capacity of the Earth's land mass and oceans in its calculations, all of which render the estimates meaningless.

Land Temperature Database Measurements

As stated earlier, there is no accurate historical temperature database of the Earth's land mass due to ad hoc sampling practices, bias induced in sampling procedures, adjustments to raw data that bias the data values and the practice of averaging data which renders the measurements meaningless. However, notwithstanding the foregoing, it is of some interest to review the statements made by the US National Oceanic and Atmospheric Administration ("NOAA"), as regards the analysis of their surface temperature database, to see what is being reported. NOAA is the organization charged with the maintenance and analysis of the US temperature databases.

On August 14, 2020, the US National Oceanic and Atmospheric Administration ("NOAA"), published a report entitled "Climate Change: Global Temperature." The report concluded that the average surface temperature of the Earth had increased by 2°F for the period 1880 - 2020. That is an increase of 0.014°F/yr., hardly a heat wave and well within the measurement margin of error using thermometers calibrated in 1°F or more increments. The data depicted a global cooling period from 1880 -1940, where the yearly average temperature dropped 0.3°C or 0.54°F, from the long-term average; and one from 1964-1977, where the average global temperature dropped by 0.2°C or 0.36°F. From 1880-1977, the concentration of CO_2 in the atmosphere reportedly increased from 280 ppm to 335 ppm. Finally, based on the report, the average global surface temperature peaked in 2017 and has been declining since, while the concentration of CO_2 in the atmosphere has risen to 410 ppm at the end of 2020. [24]

Therefore, the land surface temperature database maintained by the US Government entity responsible for that information depicts a *de minimis* warming of 0.014F/yr., well within the measurement margin of error. This is further confirmation that there has been no statistically significant warming of the Earth's land mass.

Stefan-Boltzmann Law

The Stefan-Boltzmann Law in physics plays an integral role in the AGW hypothesis in that climate scientists mistakenly use the law to calculate an increase in the surface temperature of the Earth from a hypothetical increase in radiant heat flux emanating from the Earth's lower atmosphere due to the re-emission of downwelling LWIR from an increase in the CO_2. concentration. Therefore, it is important to gain an understanding of the Stefan-Boltzmann Law and how it is designed to be used in thermodynamics analyses; and, how the UNIPCC mistakenly employs the formula.

Any body with a temperature greater than 0 K (absolute zero), emits radiant heat power as a function of its temperature. The Stefan-Boltzmann Law was formulated in 1879 by Austrian physicist Josef Stefan as a result of his experimental studies; [25, 26] the same law was derived in 1884 by Austrian physicist Ludwig Boltzmann from thermodynamic considerations. It states that the total radiant heat power emitted from a surface is proportional to the fourth power of its absolute temperature. If E is the radiant heat energy emitted from a unit area in one second (that is, the power from a unit area), and T is the absolute temperature (in Kelvin), then $E = \sigma T^4$, the Greek letter sigma (σ) representing the constant of proportionality, called the Stefan-Boltzmann constant. This constant has the value $5.670374419 \times 10^{-8}$ $W{\cdot}m^2/K^{-4}$. [26] Therefore, radiant heat power calculated from the Stefan-Boltzmann equation is expressed in Watts per meter squared per Kelvin; usually denoted as $W{\cdot}m^{-2}$

The Stefan-Boltzmann law applies only to blackbodies, theoretical surfaces that absorb all incident heat radiation. In reality, there are no thermodynamic blackbodies. A body that does not absorb all incident radiation (sometimes known as a grey body) emits less total energy than a black body. The emissivity of the surface of a material is its effectiveness in emitting energy as thermal radiation. Thermal radiation is electromagnetic radiation and it may include both visible radiation (light) and infrared radiation, which is not visible to the human eye. Quantitatively, emissivity is the ratio of the thermal radiation from a surface to the radiation from an ideal blackbody surface at the same temperature as given by the Stefan–Boltzmann law. The ratio varies from 0 to 1. [26]

Therefore, the formula to calculate the radiant heat energy emitted by a grey body like the Earth would be $E=\varepsilon\sigma T4$, where ε is the emissivity coefficient of the radiating body. While the Earth's surface has an estimated emissivity coefficient of less than 1.0 (about 0.9), most calculations using the Stefan-Boltzmann law calculate the value of the LWIR energy emitted by the Earth's surface assuming that the Earth is a blackbody with an emissivity coefficient near 1.0. A value of ε = 0.95 is often used. However, that is not the major problem in using the Stefan–Boltzmann law to calculate the radiative heat flux from the Earth's surface. **The major problem is that in order to calculate an increase in surface temperature from an increase in atmospheric concentration of so called "greenhouse gases," it is assumed that the Earth's surface temperature is in thermal equilibrium, which is not the case.** As stated earlier, the concept of an average equilibrium temperature of the Earth's surface is an abstraction and has no basis in the real world. It is a figment of the imagination of climate science to develop the Earth's energy budget and explain the concept of radiative or climate forcings. **The Stefan-Boltzmann Law can be used to calcu-**

late the theoretical radiant heat flux from the Earth's surface at an assumed constant temperature; but it cannot be used to calculate an increase in the Earth's actual surface temperature that results from an *assumed* increase in downwelling radiant heat flux from LWIR.

The Earth's net long wave infrared radiation (LWIR) emission from the surface depends on the surface and air temperatures, the humidity, and the aerosol content and cloud cover in the atmosphere. It is also a function of the absorbed solar flux at the surface, the sensible heat flux (heat transfer into the atmosphere without a phase change), convection heat transfer, latent heat flux of evaporation (heat transfer due to a phase change of liquid water to water vapor), water vapor concentration and any change in bulk air temperature of the air mass of the weather system.

Under full summer sun conditions, the peak solar heat flux reaching the Earth's surface can reach 1000 $W{\cdot}m^{-2}$ (not the 161 $W{\cdot}m^{-2}$ average depicted in the energy budget diagram above) and the short-term dry surface temperature of the Earth can easily exceed 50°C (122°F).[27] Based on the Stefan-Boltzmann law calculations, the increase in blackbody LWIR heat flux as the surface temperature increases from 20°C to 50°C is only 200 $W{\cdot}m^{-2}$ once the effect of the downward LWIR flux from the troposphere (surface energy exchange) is included. This means that most of the solar flux (~800 $W{\cdot}m^{-2}$) is coupled back into the atmosphere by convection heat transfer, not thermal (LWIR) radiation. Over the oceans, wind driven evaporation is usually the dominant cooling process, not LWIR emission. [27]

Finally, the net LWIR heat flux that is emitted from the Earth's surface is determined by the balance between the upward LWIR flux from the surface and the downward LWIR flux from the atmosphere. Under clear sky conditions, when the surface and air temperatures are similar, the net surface LWIR cooling flux is approximately 40 $W \cdot m^{-2}$. Under low humidity conditions, the net surface LWIR cooling may increase to 100 $W \cdot m^{-2}$. Under low-lying cloud cover or fog conditions, the LWIR flux may be zero. [28] There is no climate equilibrium.

It is important to understand that the equilibrium average surface temperature that is calculated by climate scientists using radiative forcing is not a measurable climate variable. There is no equilibrium temperature for the Earth. There is a significant lag between the peak solar flux and peak surface temperature due to the Earth's thermal (heat) capacity; that is, the ability of the Earth to store heat below the surface and release it later in the day. [29]

The AGW hypothesis is based on a flawed analysis of the energy flows that occur at the Earth's surface which produce energy transfer estimates within the Earth's energy budget that are incorrect. There is no climate equilibrium on any time scale and all the energy transfer processes are dynamic, not static.

CHAPTER NINE

THE THERMODYNAMIC INTERACTIONS WITH THE EARTH'S ATMOSPHERE

THE ANALYTICAL COMPLEXITY of a thermodynamic system is a function of the number of thermodynamic interactions that occur within the system or that cross the boundaries of that system. Given that definition, the Earth's atmosphere is the most complex thermodynamic system in the biosphere to analyze with respect to climate science because it interacts with the land mass, oceans, and outer space, and is impacted by all the heat transfer mechanisms at work within each.

The Sun's incoming radiation leaves outer space and enters the top of the Earth's atmosphere through the layer known as the exosphere. It then proceeds through the various layers of the atmosphere as it makes its way down to the Earth's surface. Along the way, some of that energy is reflected back into space by clouds and atmospheric components such as sulphate aerosols, and volcanic ash. It is also absorbed by ozone and water vapor in the atmosphere, and some is reflected back into the atmosphere by the Earth's surface. The energy that reaches the Earth's surface is absorbed and then transferred back to the atmosphere, and into space by various heat transfer mechanisms including convection, latent heat of vaporization, sensible heat loss and long wave infrared radiation emission (LWIR). Let's examine the basic structure of the Earth's atmosphere to understand what thermodynamic interactions occur in each one of them.

Earth's Atmosphere

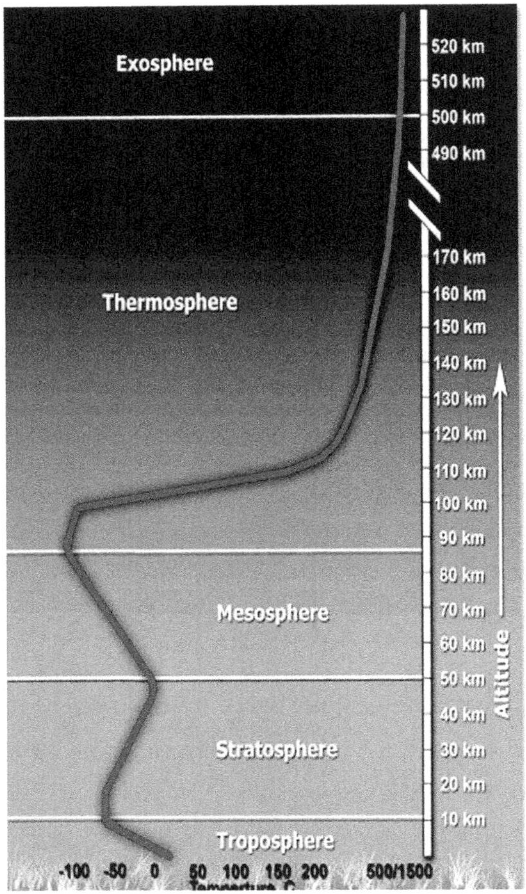

Figure 7: Climate Science Investigations: "Layers in the Atmosphere," ces.fau.edu, used by permission from NASA.

Earth's atmosphere is divided into five main layers (ascending in altitude): the troposphere, the stratosphere, the mesosphere, the thermosphere, and the exosphere. The atmosphere thins out in each higher layer until the gases dissipate into space. There is no distinct boundary between the atmosphere and space, but an imaginary line about 100 kilometers (62 miles) from the surface, called the Karman line, is usually where scientists say the atmosphere meets outer space.[1]

The troposphere is the layer closest to the Earth's surface. The thickness of the troposphere varies from about 7 to 8 km (5 mi) at the poles and to about 16 to 18 km (11 mi) at the Equator. In addition, it varies in height according to season, being lower in winter when the air is the densest. Air is warmer near the ground and gets colder with altitude in the troposphere. This phenomenon, known as the "lapse rate," results in an average 6.5°C decrease in temperature per kilometer of increase in altitude. Nearly all of the water vapor and dust in the atmosphere is in the troposphere and that is why clouds are found here.[2] The troposphere plays the major role in the determination of the Earth's climate in that almost all of the greenhouse gases are located in this layer of the atmosphere. [1]

The thermodynamic interactions within the troposphere are many in number, interrelated, interdependent and, difficult to quantify with any reasonable degree of accuracy. Most of the scientific investigations and analyses of the troposphere associated with climate science are theoretical in nature. They are based on computer models which effect changes in certain climate parameters and then attempt to predict a change in the temperature of the troposphere. In scientific investigations, it is always preferable to conduct experiments on a control volume which replicates the system under investigation. The control variables in the system under investigation are held constant, then the independent variable within the system is changed to measure the resultant effect on the dependent variable(s) within the system. In the AGW hypothesis, a change in the concentration of CO_2 in the troposphere is alleged to cause changes in other variables in the troposphere, most notably water vapor concentration and temperature. Therefore, to conduct a definitive evaluation of the hypothesis, a control volume replicating the Earth's troposphere must be created, and an experiment conducted where the control variables are held constant, the concentration of CO_2 is changed and the value of the dependent variables, temperature, and water vapor concentration, within the control volume are measured. However, it is difficult, if not impossible, to construct an accurate physical model of such a complex, dynamic, thermodynamic system such as the Earth's troposphere. As stated earlier, all of the thermodynamic interactions in the troposphere are interrelated; that is, a change in one causes a change in others, the outcomes of which are not possible to predict. Therefore, at present, climate scientists are unable to objectively test the AGW hypothesis.

The stratosphere is the second major layer of the Earth's atmosphere. The stratosphere is found from 11 to 48 kilometers (about 7 to 30 miles) above the Earth's surface. The lower stratosphere is centered around 18 kilometers (11 miles) above Earth's surface. The stratosphere is stratified (layered) in temperature, with warmer layers higher and cooler layers closer to the Earth. The increase in temperature with altitude is a result of the absorption of the Sun's ultraviolet radiation by the ozone layer. This is in contrast to the troposphere, near the Earth's surface, where temperature decreases with altitude. The border between the troposphere and stratosphere, the tropopause, marks where this temperature inversion begins. [1]

Near the equator, the lower edge of the stratosphere is as high as 20 km (12.4 miles), around 10 km (6.2 miles) at mid-latitudes, and at about 7 km (4.3 miles) at the poles. Temperatures range from an average of –51°C (–64°F) near the tropopause to an average of –15°C (5.0°F) near the mesosphere. Temperatures also vary within the stratosphere as the seasons change, reaching particularly low temperatures in the polar night (winter).

Figure 8. Space Shuttle Endeavour appears to straddle the stratosphere and mesosphere in this photo. The orange layer is the troposphere, where all of the weather and clouds, which we typically watch and experience, are generated and contained. This orange layer gives way to the whitish stratosphere and then into the mesosphere. Photo used courtesy of NASA.

Winds in the stratosphere can far exceed those in the troposphere, reaching near 134 mph (216 km/h) in the Southern polar vortex. [1]

The mesosphere starts at 50 km (31 miles) and extends to 85 km (53 miles). The top of the mesosphere, called the mesopause, is the coldest part of Earth's atmosphere, with temperatures averaging about -90°C (-130°F). Not much is known about the characteristics of the mesosphere since it is difficult to study due to its altitude and the limited technology available to do so. [1]

The thermosphere extends from 85 km (53 miles) to 500 km (310 miles) above the Earth's surface. The temperature in the thermosphere begins to increase with altitude, partly as the result of the absorption of UV and X-ray radiation; and partly due to the impact of solar wind, which is a continuous stream of protons and electrons emitted by the Sun. [1]

The exosphere represents the outermost layer of the Earth's atmosphere. It extends from 500 km (310 miles) to outer space. In this layer of the Earth's atmosphere, atoms and molecules escape into space due to the low gravitational force present at this altitude. [1]

The Role of Greenhouse Gases in the Earth's Climate

Since greenhouse gases ("GHG") play an integral role in the AGW hypothesis, let's take a more in-depth look at these important components of the Earth's atmosphere. Earth's atmosphere is comprised of many gases, most of which are in such small concentrations that they have no real effect on life on Earth. The four major gases that comprise the Earth's atmosphere, along with the percentage of their volume and concentration, are listed in Table 1. [2]

Gas Name	Chemical Formula	Percent Volume	Concentration (PPM)
Nitrogen	**N_2**	**78.08**	**780,840**
Oxygen	**O_2**	**20.95**	209,500
Argon	**Ar**	**0.93**	**9,300**
Carbon Dioxide	**CO_2**	**0.041**	**410**

Table 1. Major Gas Volume and Concentration in the Earth's Atmosphere

In case you are a Superman fan, you might like to know that Krypton (Kr) has a concentration of 1.14 ppm in the Earth's atmosphere! While water vapor (H_2O) is considered to be one of the most important gases in the Earth's atmosphere, it is not included in the list above because its concentration in the Earth's atmosphere can vary significantly in various locales. Scientists refer to water vapor and carbon dioxide as "variable gases," since their concentration in the atmosphere can vary over time depending on several factors. The concentration of CO_2 in the atmosphere can be affected by anthropogenic emissions, volcanic events (both under the sea and on land), and fires. Changes in oceanic events such as the thermocline circulation, as well as carbon dioxide exchanges within the atmosphere and ocean (carbon cycle), can also influence the atmosphere's CO_2. Photosynthesis, which is the natural phenomena involving plant matter absorption of CO_2 and the release of O_2, also plays a large role in affecting the concentration of CO_2 in the atmosphere. The concentration of water vapor in the atmosphere is affected by local weather conditions, since water is constantly cycling through the atmosphere as the result of evaporation from the Earth's surface, condensation in the lower troposphere and wind currents that affect cloud transport.

Among all of the gases present in Earth's atmosphere, five are generally classified as greenhouse gases because of their ability to absorb and radiate heat energy in the form of LWIR. Water vapor and CO_2 in our atmosphere are key to the survival of the animal kingdom on Earth due to the moderating effect that water vapor has on the Earth's near-surface temperature and the central role that CO_2 plays in plant photosynthesis. As can be seen from the table below, water vapor is the most abundant and "effective" GHG in the moderation of the Earth's temperature, estimated to contribute as much as 85% of the

GHG effect. The GHG's absorb radiant heat transfer in the form of LWIR photons from the Earth's surface and then transfer that heat to the lower troposphere by molecular collision with nearby molecules, later emitting some of that heat gained as LWIR back to the Earth's surface. This thermodynamic interaction retards some of the planet's radiant heat loss that would otherwise escape from the surface to the atmosphere and out to space (this is known as the "greenhouse gas effect"). In fact, without the greenhouse gas effect, the Earth's temperature would be much colder. There is no question that the presence of water vapor in the lower troposphere contributes to a more habitable climate on Earth.

There are four major chemical compounds that are designated as greenhouse gases. Listed below is each along with the percentage of its estimated contribution to the "greenhouse gas effect":

Greenhouse Gas	Estimated Effect (%)
Water Vapor	60-85
Carbon Dioxide (CO_2)	9-26
Nitrous Oxide (N_2O)	0.95
Methane (CH_4)	0.36
Misc. gases	0.07

Table 2. The Estimated Contribution of Greenhouse Gases to the Greenhouse Gas Effect. [2]

The concentration of water vapor in the atmosphere varies as a function of the temperature and density of the air and ranges from 0-4%. It should be noted from Table 2 that while the concentration of water in the Earth's atmosphere is estimated to average 2.5% or 25,000 ppm by volume, it is estimated to control 60% to 85% of the GHG effect. This is because it is the dominant GHG by volume and it absorbs LWIR in a broader range of the infrared radiation spectrum. Likewise, while CO_2 only occupies 0.041% of the volume of the atmosphere with a concentration of 410 ppm, it is estimated to control 9% to 26% of the GHG effect as the second largest GHG by volume. Water vapor and CO_2 are the only two gases of any significant concentration in the Earth's atmosphere that can absorb and emit LWIR. [2]

Therefore, the logical conclusion based on atmospheric concentration and LWIR absorption spectrum is that water vapor is the primary gas responsible for the GHG effect. Advocates and opponents of the AGW theory seem to agree with that conclusion. However, there is more to the story. It has to do with the purported interrelationship between water vapor and CO_2 in the warming of the Earth's troposphere.

The fact that water vapor is proven to be the most "potent" and effective GHG is widely recognized by the climate science community based on its absorption spectrum and concentration in the atmosphere. The fact that the concentration of water vapor in the atmosphere is the result of natural causation and man's activities have virtually no effect on that level of concentration (other than the purported water vapor feedback cycle theory employing CO_2) is not lost on AGW proponents as well. Therefore, the hypothesis that has been advanced by AGW proponents that links man to global warming is directly related to the purported link between CO_2 in the atmosphere and water vapor concentration.

The Anthropogenic Global Warming Hypothesis

The AGW hypothesis is primarily concerned with what happens to the LWIR that is emitted by the Earth's surface into the lower troposphere. Some of that LWIR heat energy is absorbed by the molecules of gases in the atmosphere known as "greenhouse gases." Some of it passes through the "atmospheric transmission window" and escapes into space. The term "greenhouse gas" is really a misnomer and is wrongly attributed to Joseph Fourier, a French mathematician and physicist. [4]

Fourier's greatest contributions to science occurred in the fields of mathematics and thermodynamics, primarily dealing with developing the mathematical equations to quantify the conduction heat transfer in solid bodies. In 1822, Fourier published *Théorie analytique de la chaleur* (The Analytical Theory of Heat), in which he based his theory regarding the flow of heat between two adjacent molecules in a solid body on Newton's law of cooling. Fourier also developed the mathematical framework with which to quantify the rate of heat transfer within a solid body, known as the heat equation, which is a partial differential equation used today to quantify the rate of conductive diffusion. [4]

In the 1820s, Fourier calculated that an object the size of the Earth, at its distance from the Sun, should be considerably colder than the planet is if warmed only by the effects of incoming solar radiation. He examined various possible sources of the additional observed heat in articles published in 1824 and 1827. While he ultimately suggested that interstellar radiation might be responsible for a large portion of the additional warmth, Fourier's consideration of the possibility that the Earth's atmosphere might act as an insulator of some kind is widely recognized as the first proposal of what is now known as the greenhouse effect, although Fourier never called it that.[4]

In his articles, Fourier referred to an experiment by de Saussure, who lined a vase with blackened cork. Into the cork, he inserted several panes of transparent glass, separated by intervals of air. Midday sunlight was allowed to enter at the top of the vase through the glass panes. The temperature became more elevated in the more interior compartments of this device. Based on de Saussure's experiment, Fourier concluded that gases in the atmosphere could form a stable barrier like the glass panes. This conclusion may have contributed to the later use of the metaphor of the "greenhouse effect" to refer to the processes that determine atmospheric temperatures. Fourier noted that the actual mechanisms that determine the temperatures of the atmosphere included convection, which was not present in de Saussure's experimental device. [4]

As described above, de Saussure's model, using an enclosed "greenhouse" structure, eliminated one of the most important heat transfer mechanisms at work in the atmosphere's thermodynamic interactions: convection heat transfer at the Earth's surface.

Next came Svante Arrhenius, a Swedish physicist and Nobel prize winner for his work in chemistry in 1903. Arrhenius was the father of the modern-day theory of "radiative forcing" in that he reasoned that an increase in CO_2 concentration in the troposphere could result in an increase in the temperature of the troposphere which would cause an increase in water vapor, the most effective greenhouse gas. However, it should be noted that Arrhenius omitted the thermodynamic effects of clouds, convection of heat upward in the atmosphere, and other essential factors. [5]

Arrhenius was not knowledgeable about the history of CO_2 emissions into the atmosphere, so he turned to a Swedish colleague, Arvid Högbom, for assistance. Högbom had compiled estimates for how carbon dioxide cycles through natural geochemical processes, including emission from volcanoes, uptake by the oceans, and so forth. In addition, he had calculated the anthropogenic contributions to the concentration of CO_2 into the atmosphere as of the late 1800s. Arrhenius calculated that doubling the CO_2 concentration would cause the Earth's temperature to increase 4°C. Interestingly, Arrhenius believed that an increase in the global temperature would be beneficial for mankind in that it "would warm cooler climates and enhance agricultural production." [6] Little did he know that over 100 years later, climate scientists promoting the AGW hypothesis would predict catastrophic consequences from what Arrhenius believed to be a beneficial effect. However, neither he nor Högbom were concerned about that prospect, given that at the then current rate of anthropogenic emissions in 1896, they estimated that it would take 3,000 years to double the CO_2 concentration in the atmosphere. As the rate of CO_2 emissions began to rise at an increasing rate, succeeding climate scientists began to sound the CO_2 global warming alarm. [7]

The proponents of the AGW hypothesis postulate that the LWIR that is emitted by the Earth's surface is "trapped" (absorbed) by the CO_2 molecules in the Earth's troposphere and then immediately re-radiated back to the Earth's surface, causing the surface temperature to increase. The theory is that the more the CO_2 concentration increases in the atmosphere, the more LWIR is "trapped" by CO_2 molecules and reradiated back to Earth, causing further warming.

In its AR5 Report in 2013, the UN IPCC states the following concerning radiative forcings (RF): [8] *"RF is the net change in the energy balance of the Earth system due to some imposed perturbation. It is usually expressed in watts per square meter averaged over a particular period of time and quantifies the energy imbalance that occurs when the imposed change takes place.* ***Though usually difficult to observe*** *(emphasis added), calculated RF provides a simple quantitative basis for comparing some aspects of the potential climate response to different imposed agents, especially global mean temperature, and hence is widely used in the scientific community. Forcing is often presented as the value due to changes between two particular times, such as pre-industrial to present-day, while its time evolution provides a more complete picture."*

I would interpret the statement above as follows: Radiative forcing is the net change in the theoretical energy balance of the Earth's system due to an imposed change in one of the atmospheric parameters (such as the concentration of CO_2, water vapor, etc.) in the calculations of a climate model. The value of the theoretical energy change is expressed in $W{\cdot}m^{-2}$ over a specified time period. The concept of radiative forcing is theoretical and has not been validated by physical observation in nature. However, it provides climate scientists with a simple way to calculate prospective changes in the global mean temperature of the Earth when imposing changes in the values of atmospheric parameters (such as CO_2 and water vapor). Since all climate scientists employ this technique in our climate model calculations, it must be the right thing to do. The assumptions involved in this process simplify the resultant calculations. The theoretical concept of radiative forcings is especially useful to compare the climate parameters and estimated global mean temperature that existed before man began to emit CO_2 into the environment with present values of those same parameters.

The AR5 report further states: *"Climate change takes place when the system responds in order to counteract the flux changes, and all such responses are explicitly excluded from this definition of forcing. The assumed* *relation between a sustained RF and the equilibrium global mean surface temperature response (ΔT) is $\Delta T = \lambda RF$ where λ is the climate sensitivity parameter. The relationship between RF and ΔT is an expression of the energy balance of the climate system and a simple reminder that the steady-state global mean climate response to a given forcing is determined both by the forcing and the responses inherent in λ." [8]*

I would interpret the above as follows: climate change occurs when nature responds to counteract the effect of energy changes in the atmosphere due to radiative forcings. However, we excluded those responses from our definition of forcing. The assumed relationship between a radiative forcing that continues for an extended period and the equilibrium global mean surface temperature is a function of the climate sensitivity factor multiplied times the value of the radiative forcing.

We assume that the effect (change) on the equilibrium global mean surface temperature is a function of the theoretical climate sensitivity parameter Lambda, multiplied by the value assigned to the radiative forcing RF. The climate sensitivity parameter Lambda quantitatively represents how strongly the climate reacts to the imposed change in a climate parameter like CO_2. The relationship between the radiative forcing and the change in equilibrium global mean surface temperature is an expression of the energy balance of the climate system and how changes in that energy balance affect temperature. The global mean climate response to a given change in a climate parameter is determined by the nature of the forcing and the response that we build into the climate sensitivity factor.

More from AR5: *"Implicit in the concept of RF is the proposition that the change in net irradiance in response to the imposed forcing alone can be separated from all subsequent responses to the forcing. These are not in fact always clearly separable and thus* ***some ambiguity exists in what may be considered a forcing versus what is part of the climate response*** *(emphasis added)." [8]*

I find this statement by the UNIPCC to be particularly ironic, given that the climate models are filled with unproven assumptions about the relationships among climate parameters such as air temperature, density, relative humidity, CO_2. and water vapor concentration and how a change in one might affect the others. To state that the model predictions about a change in downwelling net heat flux that results from a "forced" change in a model climate parameter such CO_2. concentration might not correlate with actual results, states the obvious.

I don't normally consider the role of a climate scientist to be creative, *per se.* Their job should be to observe phenomena, postulate a hypothesis to explain why such phenomena occur and then conduct investigations (experiments, research, etc.) to falsify the hypothesis. If the hypothesis cannot be falsified, then it rises to the level of a theory. However, someone must have channeled the spirit of Svante Arrhenius when they came up with the idea of a radiative forcing. If "necessity is the mother of invention," the concept of radiative forcing was born of necessity. It is the underpinning of the global warming hypothesis.

The global warming hypothesis rests mainly on the theoretical concept of radiative forcings and feedbacks, and it goes like this: an increased concentration of CO_2 molecules in the Earth's atmosphere due to anthropogenic emissions of CO_2 results in the added CO_2 molecules "trapping" (absorbing) more LWIR emissions from the Earth's surface. The subsequent downward (towards the Earth's surface) reradiating of this "trapped" LWIR by the CO_2 molecules increases the temperature of the

Earth's surface, which in turn causes more LWIR to be emitted, which gets absorbed by CO_2 and water vapor. As a result, the temperature of the lower troposphere increases, and more water vapor is formed in the troposphere. Since water vapor is acknowledged to be the most potent greenhouse gas, the increased water vapor absorbs additional LWIR that is emitted by the Earth's surface, thereby creating what is known in climate science as a "positive feedback." A positive feedback increases an initial warming effect. A positive feedback that is self-perpetuating, such as the example above, is known as a "positive feedback loop." [10, 11]

Climate scientists have identified a number of positive feedback loops in the climate system. One example is melting ice. Because ice is light-colored and reflective, a large proportion of the sunlight that hits it is reflected back to space, which limits the amount of warming it causes. But as the world gets hotter, ice melts, revealing the darker-colored land or water below. The result is that more of the Sun's energy is absorbed, leading to more surface warming, which in turn leads to more ice melting – and so on. A climate scientist is limited only by his or her imagination as to how much additional warming is created by positive feedbacks. Since it is all theoretical in nature, the sky is the limit (pardon the pun) or at least, the troposphere. [10, 11]

As a result of the radiative forcing and feedback loop concepts, many climate scientists today mistakenly believe that the emission and absorption of LWIR by the Earth's surface and greenhouse gases, and CO_2 in particular, is the predominate heat transfer mechanism to cool (and heat) the Earth's surface. This mistaken view is due to what appears to be a lack of understanding of spectroscopy, quantum mechanics and the heat transfer interactions that occur at the Earth's surface and within the first kilometer of the troposphere.

At 1 km in altitude, H_2O absorbs 75% of the LWIR heat flux emitted by the Earth's surface (~60 $W{\cdot}m^{-2}$) and CO_2 absorbs about 25% of the LWIR heat flux emitted by the Earth's surface (~20 $W{\cdot}m^{-2}$). At 2 m (6.5 ft.) above the Earth's surface, H_2O absorbs 90% of the LWIR heat flux emitted by the Earth's surface (14.4 $W{\cdot}m^{-2}$) and CO_2 only 10% of the LWIR heat flux emitted by the Earth's surface (1.6 $W{\cdot}m^{-2}$). [9] This data was derived from the spectrally resolved fraction of the LWIR flux with a surface temperature set to 325 K (125°F), air temperature of 295 K (71°F), CO_2~380 ppm and RH = 50%, using a high resolution (0.01cm-1) radiative heat transfer model.[12] The heat power radiated by the Earth's surface at a temperature of 325 K is 632 $W{\cdot}m^{-2}$ based on the Stefan-Boltzmann formula. [10, 11]

The above calculations clearly demonstrate that the maximum heat flux from LWIR emissions from the Earth's surface is around 80 $W{\cdot}m^{-2}$ at 1 km. Yet, the total heat flux is 632 $W{\cdot}m^{-2}$. Therefore, other heat transfer mechanisms, primarily convection heat transfer, are responsible for more than 80% of the total heat transfer involved in cooling the Earth's surface. [10, 11]

In reality, it is convection heat transfer at the Earth's surface; water vapor concentration in the troposphere; and molecular collision within the first kilometer of the atmosphere that are predominately responsible for regulating the temperature of the lower troposphere. *CO_2 plays virtually no role in the Earth's climate. Water vapor is the dominant greenhouse gas, and its concentration is unaffected by CO_2 concentration. [11]*

Let's examine the statement above to develop a better understanding of what actually takes place from a thermodynamic standpoint in the

Earth's relevant atmosphere (the lower troposphere) to moderate the Earth's temperature.

Convection Heat Transfer at the Earth's surface

As was discussed in Chapter 8, there are many factors that affect the temperature of the Earth's surface at a given latitude and point in time. The solar flux changes with both the seasons and time of day. These changes are related to the ellipticity of the Earth's orbit, the axial tilt, and the daily rotation. The reflected solar flux changes with the composition of the Earth's surface (albedo effects). Local weather conditions affect the rate of convection heat transfer, surface air temperature and humidity; and atmospheric conditions in the lower troposphere affect relative humidity and cloud cover.

Under full summer sun conditions, the peak solar heat flux reaching the Earth's surface can reach 1000 $W{\cdot}m^{-2}$, not the 161 $W{\cdot}m^{-2}$ depicted in the "average" energy budget diagram in Chapter 8. The short-term dry surface temperature of the Earth can easily exceed 50°C (122°F). [3] Based on the Stefan-Boltzmann law calculations, the increase in blackbody LWIR heat flux as the surface temperature increases from 20°C (68°F) to 50°C (122°F) is 200 $W{\cdot}m^{-2}$. This means that most of the solar flux (~800 $W{\cdot}m^{-2}$) is coupled back into the atmosphere by convection heat transfer, not thermal (LWIR) radiation. [11] **It is not physically possible for the Earth to lose all of the heat it gains from solar irradiance by the emission of LWIR,** as noted above. Convection heat transfer is responsible for up to 80% of the heat loss during the day; emission of LWIR for about 20%. However, when the local surface temperature is approximately equal to the temperature of the air above the surface, LWIR emission becomes the predominant heat transfer mechanism. Let's examine what happens to the photons that are emitted in this process.

Spectroscopy

Spectroscopy is the branch of science that is concerned with the investigation and measurement of spectra produced when matter absorbs or emits electromagnetic radiation.[13] A spectrum is defined as a characteristic series of frequencies (or wavelengths) of electromagnetic radiation emitted or absorbed by a substance. Current spectroscopic analysis often uses the terminology "wave numbers" to describe the length of a wave and that designation is denominated in reciprocal centimeters, or cm^{-1}. A wave number is the reciprocal of the wavelength expressed in meters multiplied by a constant, 10,000. Historically, early research in spectroscopy described the wavelength in micrometers, or μm, which is 10^{-6} meters; or, nanometers, nm, which is 10^{-9} meters. Wavelengths in the mid-infrared range are often expressed in micrometers, commonly called microns, where one micron is 10^{-6} meters. The wave number is proportional to wavelength based on the following formula: [14]
Wave Number cm^{-1} = (10,000) / (Wavelength microns)

For example, to find the wave number for a wavelength = 15 μm

$$\text{Wave Number} = (10{,}000)/15 = 666.66\ cm^{-1}, \text{ or } 667\ cm^{-1}$$

If the wavelength is 550 nm (550×10^{-9} m), then

$$WN = (10{,}000) / (0.555 \times 10^{-6}\ m) = 18018\ cm^{-1}$$

Both wavelengths expressed in reciprocal centimeters (cm^{-1}) and wavelength expressed in microns (μm) will be used intermittently to describe LWIR wavelength in this chapter, as appropriate.

Spectral line strength denotes the ability of a molecule to absorb photons of that wavelength and transition to a higher energy state. The energy of a photon emitted at a spectral line is calculated using the Planck Equation, or, Planck-Einstein Relation, E = hv, where E = energy, h = Planck's constant and v = frequency of the wave. The Planck-Einstein relation is a formula integral to quantum mechanics; it states that a quantum of energy (E) is equal to the Planck constant (h) = 6.62×10^{-34} J, times a frequency of oscillation of an atomic oscillator (ν, the Greek letter nu). It is used to calculate the energy of a photon at a specified wavelength. For example, a photon with a wavelength of 15 microns has an energy of 0.0827 eV (electron volts), or 1.325×10^{-20} joules. [14] Therefore, using the Planck-Einstein Relation Equation, one can calculate the energy absorbed/emitted by CO_2 in the atmosphere based on the wavelength of the absorbed photon.

Quantum Mechanics

"Anyone who is not shocked by quantum theory has not understood it."
- Neils Bohr [15]

"If it is correct, it signifies the end of physics as a science."
- Albert Einstein [15]

"I think I can safely say that nobody understands quantum mechanics." - Dr. Richard Feyman, an American theoretical physicist who won the 1965 Nobel Prize for his work in the field of quantum electrodynamics. [15]

Quantum mechanics is the branch of physics that deals with the mathematical description of the motion and interaction of subatomic particles, incorporating the concepts of quantization of energy, wave-particle duality, the uncertainty principle, and the correspondence principle. [12, 16] The field of quantum mechanics plays an important role in understanding what occurs when a molecule absorbs a photon of LWIR and the resultant effect on an electron within that molecule and the molecular bond.

The field of quantum mechanics proposes an alternate way of viewing electromagnetic radiation (EMR). That is, that EMR consists of photons, uncharged elementary particles with zero rest mass, which possess discrete quantities of energy, or quanta, of the electromagnetic force which is responsible for all electromagnetic interactions. These photons exhibit a wave-particle duality; that is, the exhibition of both wavelike and particle - like properties by a single entity. For example, photons undergo diffraction and can interfere with each other as waves, but they also act as point-like particles that have momentum, but not mass. [12]

Quantum electrodynamics is the theory of how EMR interacts with matter on an atomic level. Quantum effects provide additional sources of EMR, such as the transition of electrons to higher or lower energy levels in an atom. The energy of an individual photon is quantized and is greater for photons of higher frequency (shorter wavelength). A single gamma ray (shortest wave-length radiation) photon, for example, might carry 60,000 ~ 100,000 times the energy of a single photon of visible light. [12]

The field of quantum mechanics is not without its controversy. In fact, one could argue that it was born of controversy. Max Planck is often called the father of quantum mechanics. Yet, it was Niels Henrik David Bohr, a Danish physicist, who made foundational contributions to understanding atomic structure and quantum theory. For this achievement, he received the Nobel Prize in Physics in 1922. [17]

Figure 9. "The Fifth Solvay Conference, probably the most intelligent picture ever taken, 1927" (17 of the 29 attendees were or became Nobel Prize winners.) From right to left: Back: Auguste Piccard, Émile Henriot, Paul Ehrenfest, Édouard Herzen, Théophile de Donder, Erwin Schrödinger, JE Verschaffelt, Wolfgang Pauli, Werner Heisenberg, Ralph Fowler, Léon Brillouin. Middle: Peter Debye, Martin Knudsen, William Lawrence Bragg, Hendrik Anthony Kramers, Paul Dirac, Arthur Compton, Louis de Broglie, Max Born, Niels Bohr. Front: Irving Langmuir, Max Planck, Marie Curie, Hendrik Lorentz, Albert Einstein, Paul Langevin, Charles-Eugène Guye, C.T.R. Wilson, Owen Richardson. Photo courtesy of the Solvay Library.

The 5th Solvay Conference in 1927 saw Einstein debating with attendees Niels Bohr and Werner Heisenberg about how quantum mechanics had gone too far and reduced the behavior of subatomic particles to probabilities. It was the Heisenberg Uncertainty Principle or Indeterminacy Principle, articulated by the German physicist Werner Heisenberg in 1927, which caused the controversy. Heisenberg proposed that the position and the momentum of an electron cannot both be measured exactly at the same time, even in theory. In fact, he argued that the concepts of exact position and exact velocity together have no meaning in nature. [18]

Einstein, being trained in classical Newtonian physics, found it difficult to accept the probability theories of Heisenberg that were embedded in the quantum theory. The problem lies with determining the position of an electron within an atom. The Uncertainty Principle states that the square of the wave function, Ψ^2, represents the probability of finding an electron in a given region within the atom. An atomic orbital is defined as the region within an atom that encloses where the electron is likely to be 90% of the time. [18] While Einstein came to embrace quantum theory in his later years, he remained troubled by the Uncertainty Principle. While some aspects of quantum theory remain unresolved, the science regarding the quantum effects of photon absorption by an electron seems to be widely accepted.

So, why is an understanding of spectroscopy and quantum mechanics important in understanding the transfer of heat in the troposphere from the absorption and emission of LWIR photons by a CO_2 molecule? It has to do with what happens when the electrons within those molecules absorb the quanta of energy carried by the LWIR photons emitted from the Earth. Does that energy really get "trapped" and get immediately reradiated back to Earth as downward LWIR heat flux from the lower troposphere, heating the Earth's surface as the UN IPCC and global warming proponents suggest? Is the value of that heat flux of such a magnitude that it actually creates global warming? Is the value of that downward heat flux of such a value that it results in the heating of the lower troposphere and causes an increase in water vapor concentration, resulting in the absorption of more LWIR and a feedback loop? Or are there other heat transfer mechanisms at work that falsify that hypothesis?

Since CO_2 and H_2O are the dominant greenhouse gases, let's examine the

LWIR absorption spectrum for each molecule to determine what happens from a quantum mechanics perspective when the electrons within those molecules absorb the energy in the LWIR photons emitted by the Earth.

Photon Absorption in the Long Wave Infrared Spectrum

There are three main processes by which a molecule can absorb the energy in a LWIR photon emitted by the Earth's surface. Each of these processes involves an increase in energy within the absorption molecule that is proportional to the energy of the photon absorbed. The first process occurs when absorption of photon energy leads to a *higher rotational* energy level within the molecule in a rotational transition. The second process is a vibrational transition which occurs on absorption of quantized energy, which leads to an *increased vibrational* energy level. The third process involves electrons of molecules being raised to a higher orbital level, in what is known as an *electronic transition.* It is important to state that the energy is quantized (a discrete amount based on the wavelength), and absorption of radiation causes an electron to move to a higher internal energy level. There are multiple possibilities for the different possible energy levels for the various types of transitions. The two major molecular vibrational modes that can occur in a molecule include *stretching that changes the bond length* and *bending* that changes the bond angle. [19, 20]

The normal vibrational modes in a water vapor molecule include the symmetric stretch, the asymmetric stretch and bending. The water vapor molecule has three degrees of vibrational and rotational freedom. Figure 10 shows the three possible vibrational modes of water vapor and CO_2. The vibrational modes for H_2O are labeled with v1 (symmetrical stretch), v2 (bending) and v3 (asymmetrical stretch). The figure also shows the wave number and wavelength associated with the absorption at band center. [19, 20]

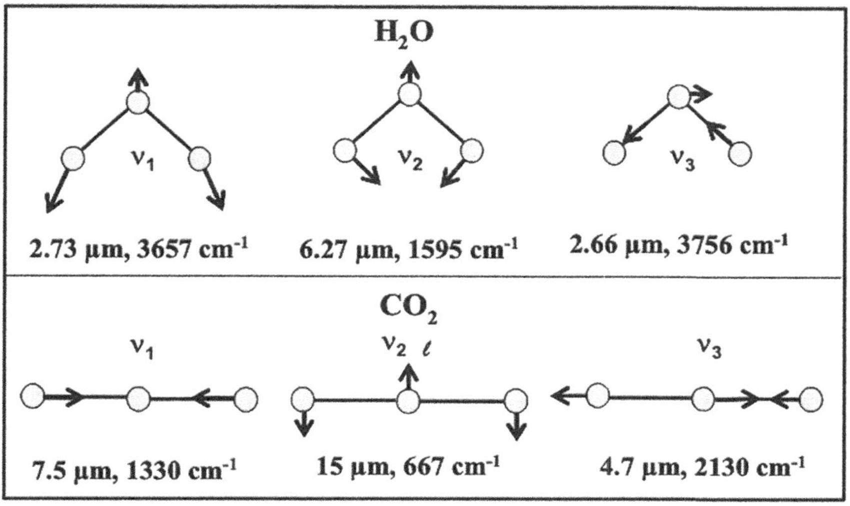

Figure 10. The Normal Modes of Vibration of Water Vapor and CO_2 **Molecules.**

Carbon dioxide, a linear molecule, has three normal modes of vibration. Even though it does not have a permanent dipole moment, the dipole moment changes during two of the three modes; thus, fulfilling the quantum requirement that a radiative molecule have a dipole moment. Carbon dioxide can absorb photons in the LWIR spectrum in two of its normal modes: asymmetrical stretch and bending. Figure 10 depicts the three normal vibrational modes of CO_2; they are labeled with v1 (symmetrical stretch), v2 (bending) and v3 (asymmetrical stretch). [21]

Each one of the absorption bands depicted for each molecule contains many spectral lines that are transitions between the rotational energy levels of the molecule in the upper and lower energy state.

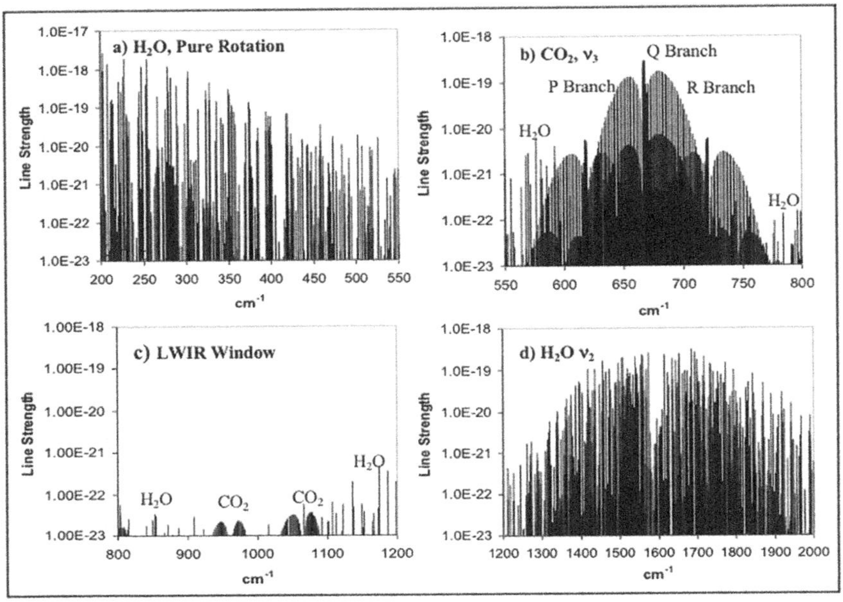

Figure 11. Line Strength of CO_2 and H_2O from 200-2000 cm^{-1}. Courtesy of Roy Clark, Ph.D. [9]

Figure 11 depicts the line strength and spectral lines within the absorption bands for CO_2 and H_2O in the IR spectrum as determined by high resolution spectroscopy (0.01 cm^{-1}). To be absorbed by either CO_2 or H_2O, a photon must have a wave number (~ wavelength) that falls within these spectral bands. The higher the wave number (cm^{-1}), the higher the energy level of the photon absorbed. The principal bands for H_2O are the pure rotational band, a), which lies principally with 200 to 550 cm^{-1}, and the lowest vibrational transition spectrum for H_2O, d), which lies within the range of 1200 cm^{-1} to 2000 cm^{-1}. The principal absorption band for CO_2 is the lowest fundamental vibration transition, b), which lies within the range of 550 cm^{-1} to 800 cm^{-1}. [9]

Figure 11 also clearly demonstrates that water vapor has a broader absorption spectrum and absorbs photons of higher and lower energies than does CO_2. It is this characteristic, along with its concentration level within the Earth's atmosphere, that makes H_2O the dominant greenhouse gas, not CO_2. The absorption spectrum of CO_2 consists of many overlapping lines with H_2O, including those in the 15-micron (667 cm^{-1}) region. The high-resolution spectrum can be calculated using the spectral data available from the publicly available HITRAN database. The graph subtitled "c) LWIR Window" depicts the spectrum for the LWIR transmission window, which lies in the 800 to 1200 cm^{-1} spectral range. As one can see, neither H_2O nor CO_2 absorb much photon energy within these spectral lines. Some of the LWIR emission from the Earth within the 800 to 1200 $cm^{-}1$ spectral range passes unabsorbed through the LWIR transmission window and into space.

The Mechanics of the Absorption and Emission of IR Photons in the Lower Troposphere

When a CO_2 molecule absorbs a LWIR photon in the lower troposphere, it gains energy in direct proportion to the wavelength of the LWIR photon absorbed, as described above. The added energy causes the molecular bond to stretch or bend, based on the energy (wavelength) of the absorbed photon. A molecule that has absorbed a photon and gained energy is referred to as "excited" since an electron within an atom in the molecule has absorbed that energy and moved to a higher energy state. However, a fundamental principle of quantum mechanics is that an excited electron "prefers" to be in a lower energy level (orbital) known as the "ground state." Therefore, the electron will not stay in the higher energy level and will seek a means to transfer the energy gained by the absorption of the photon either by molecular collision or spontaneous emission.[19]

Due to the molecular density in the lower troposphere, which is the result of the Earth's gravitational force, molecules rapidly collide with adjacent molecules in what are known as molecular collisions. The mean collision time (the time that elapses before one molecule collides with another) for an excited CO_2 molecule in the lower atmosphere is $<10^{-9}$ seconds, or one-billionth of a second. In the absence of collisions, an excited molecule will emit a photon to return to a lower energy state in a process known as spontaneous emission. The mean decay time (the time that elapses before an excited electron spontaneously emits a photon) of an excited CO_2 molecule in the lower troposphere is near one second. This means that the excited state of a CO_2 molecule is rapidly quenched by molecular collisions, mainly with N_2 or O_2 molecules. [19] What happens when an excited CO_2 molecule collides with a nearby molecular in the atmosphere in a molecular collision?

The energy gained by the absorption of the photon energy is transferred to the thermal energy of the local air mass and is not re-radiated by the CO_2 molecule. The initial increase in temperature of the air mass is simply the energy of the total number of photons absorbed by the CO_2 molecules in the local air parcel divided by the heat capacity. However, as the air warms up and expands, the air parcel may become buoyant and start a convective ascent through the troposphere, thereby transferring some of the heat gained to cooler air above it by convection. **As a result,** ***some portion of the LWIR heat energy of a photon, which is absorbed by a CO_2 molecule, is then transferred to the upper atmosphere by convection heat transfer and ultimately into space; and not as downward heat radiated flux to the Earth's surface.*** [19]

The IR active molecules in the air (mainly H_2O and CO_2) are always absorbing and emitting photons. The emission at each (monochromatic) wavelength is the blackbody emission at the temperature of the local air parcel scaled by the absorption along the observation path length. The upper limit to the emission is given by Planck's law. The emission is isotropic (equal intensity in all directions). Near the ground, the discrete molecular lines within the main absorption bands merge and form a quasi-continuum. [19, 20]

The downward long wave IR (LWIR) emission from the emission bands in the lower troposphere to the surface provides what is known as an "exchange energy." The downward photons are exchanged with those emitted by the surface. If the surface and air temperatures are the same, then there is no net LWIR energy (heat) transfer between the surface and the air within the spectral range of the main molecular bands. The conclusion from the above analysis is that only a part of the heat gained by a CO_2 molecule that absorbs a LWIR photon is re-radiated back to the Earth's surface. A portion of that heat is transferred by convection to cooler air in the troposphere. [19, 20]

The annual variation in the atmospheric CO_2 concentration for 2019 is shown in Figure 12. Over the last 200 years, the atmospheric concentration of CO_2 has increased by about 120 parts per million (ppm) from 280 to 400 ppm. Radiative transfer calculations show that the increase in CO_2 concentration of 120 ppm has produced an increase in downward LWIR flux to the surface of approximately 2 $W \cdot m^{-2}$. Similarly, a doubling of the CO_2 concentration from 400 to 800 ppm will produce an increase in downward LWIR flux of approximately 5.7 $W \cdot m^{-2}$. [22] However, this must be combined with all the other flux terms that interact with the surface layer to calculate the net heat flux.

It is the small amount of this added heat flux from the increased CO_2 concentration, when compared to the value of total solar irradiance that reaches the Earth's surface, that reveals the true role that CO_2 concentration plays in the Earth's climate.

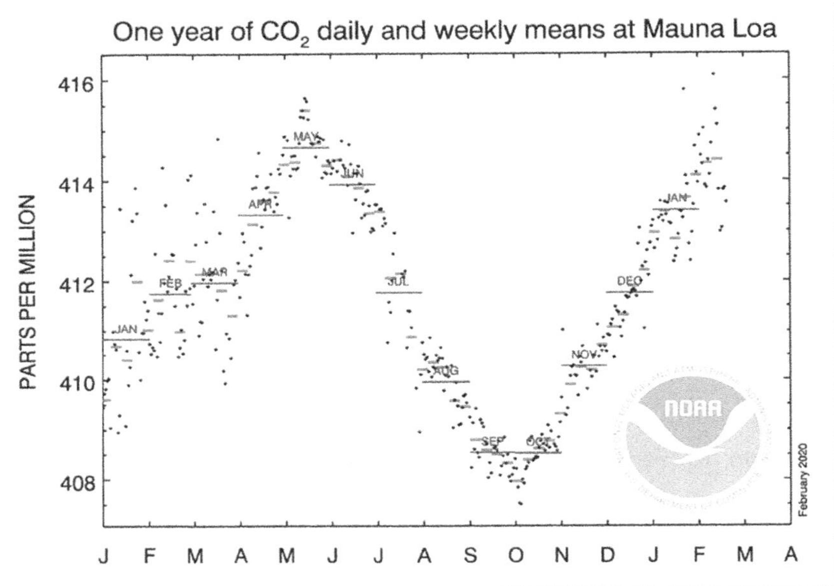

Figure 12. "One Year of CO_2 Daily and Weekly Means at Mauna Loa, April 20, 2019 – April 20, 2020." Courtesy of NOAA. [22]

Under full summer sun or tropical illumination conditions, the peak daily solar flux may easily reach 1000 $W{\cdot}m^{-2}$. The change in surface temperature is the change in heat content (enthalpy) of the surface layer divided by the heat capacity. When the time dependent surface heating and cooling is determined using thermal engineering methods, **the additional increase in surface temperature from doubling the CO_2 concentration in the atmosphere is too small to measure.** [20]

In his research titled, "A Null Hypothesis for CO_2," Dr. Roy Clark, Ph.D., calculated that an increase of 100 ppm in CO_2 concentration in the troposphere would result in an increase in the downward atmospheric LWIR heat flux of 1.7 $W{\cdot}m^{-2}$. Dr. Clark noted that his radiative transfer calculations were performed using publicly available spectrographic data from the Harvard University HITRAN database and were the same values used in the UN IPCC computer models. Dr. Clark calculated that an increase in the downward LWIR heat flux of 1.7 $W{\cdot}m^{-2}$ would result in an increase in the peak summer temperature of 0.32°C and a winter temperature of 0.34°C, with the maximum changes occurring at night. Daytime increases are less, 0.14°C and 0.17°C respectively. [11]

In an effort to simulate the ambient conditions often used in climate models and radiative heat transfer calculations concerning LWIR emissions from the Earth, Dr. Clark calculated the change in downward heat flux resulting from an increase of 100 ppm in the atmospheric concentration of CO_2 at a surface temperature of 325 K (125°F), with an air temperature of 295 K (71°F) and 50% relative humidity, at two meters above the ground. In that case, a total heat flux of 16 $W{\cdot}m^{-2}$ is absorbed, with CO_2 only absorbing 1.6 $W{\cdot}m^{-2}$, or 10%. Water Vapor absorbs the balance.[11]

The relative humidity of the atmosphere (amount of water vapor concentration) is the determining factor in the absorption of LWIR emissions from the Earth's surface and the resultant heat flux. Using the same parameters for the surface and air temperatures and increased CO_2 concentration of 100 ppm, Dr. Clark analyzed the change in the heat flux absorbed at two meters above the ground when the relative humidity was increased from 10% to 90%. The change in the total flux from an increase in RH from 10% to 90% was 12 W·m-2. Of this amount, only 0.25 W·m2 (2%) was absorbed by CO_2. [11]

The obvious conclusion from Dr. Clark's research is that the negligible amount of theoretical downward heat flux from the absorption of LWIR by CO_2 molecules in the atmosphere is dwarfed by the changes in heat flux due to the amount of water vapor in the atmosphere and changes in the solar irradiance due to all the aforementioned factors. Dr. Clark's research proves that the concentration of water vapor in the atmosphere predominates the absorption of LWIR photons emitted from the Earth's surface. Water vapor is the dominant greenhouse gas; CO_2 plays a minor role in controlling the Earth's temperature.

On May 2, 2018, Dr. B.M. Smirnov, Ph.D., published a research paper entitled, "Collision and radiative processes in emission of atmospheric carbon dioxide" in the *Journal of Applied Physics*. The paper had previously been published by the *Joint Institute for High Temperatures,* in Moscow, Russia. [23] Dr. Smirnov used a calculation methodology based on the standard atmospheric model, classical molecular spectroscopy, and the regular model of the spectroscopy absorption band. The standard atmospheric model is used by scientists to establish atmospheric parameters such as temperature, atmospheric pressure and lapse rate based on altitude, for research to provide comparability.

Dr. Smirnov describes the reasoning behind his methodology as follows:[23]

"The radiative flux from the atmosphere toward the Earth is represented as that of a blackbody and the radiative temperature for emission at a given frequency is determined with accounting for the local thermodynamic equilibrium, a small gradient of the troposphere temperature and a high

optical thickness of the troposphere for infrared radiation. Because of large radiative lifetimes for vibrationally excited molecules, they are found in thermodynamic equilibrium even at not very large pressures of a gas where radiative molecules are located."

"According to the above analysis, ***collision processes dominate in the course of propagation of infrared photons in the troposphere.*** *This means that* ***an excited molecule state resulted from absorption of a resonant photon is quenched in collisions with air molecules*** (emphasis added)." [23]

Smirnov determined that *"the rate of quenching* (rate at which molecules give up the added kinetic energy acquired through photon absorption by molecular collision) *of vibrationally excited states in the troposphere is large compared to the radiative rate."* [23] In other words, Like Dr. Clark, Dr. Smirnov finds that molecular collision, not reradiating photons, is the heat transfer mechanism that CO_2 molecules employ to transfer heat acquired through the absorption of LWIR photons.

Both scientists determined that the downwelling radiative flux towards Earth was represented as a blackbody continuum emitting photons with radiative power as a function of the Stefan-Boltzmann Law based on the temperature of the air parcel acting as a blackbody continuum.

Dr. Smirnov calculated that an increase in the atmospheric concentration of CO_2 from the current value of approximately 400 ppm to 800 ppm would result in a net increase in the downward heat flux of 6 $W{\cdot}m^{-2}$ (1.5 $W{\cdot}m^{-2}$/100 ppm), with an increase in global temperature of 0.4 K, or 0.4°C. It should be noted that Dr. Smirnov's calculation of 1.5 $W{\cdot}m^{-2}$/100 ppm is in close agreement with Dr.

Clark's calculation of 1.7 $W \cdot m^{-2}$/100 ppm. Finally, Smirnov concludes: *"The corresponding analysis convinces us that* ***contemporary injection of carbon dioxide in the atmosphere as a result of combustion of fossil fuels is not important for the greenhouse effect.*** *(Emphasis added)"* [23]

Conclusion

The predominant means of the cooling of the Earth's surface is by convection heat transfer, not the emission of LWIR radiation. A secondary means of cooling is by radiative heat transfer through the emission of LWIR from the Earth's surface, which either passes through the atmospheric transmission window or is absorbed by water vapor and CO_2 molecules. There is no equilibrium temperature for the Earth's surface. The photon energy in the LWIR emitted by the Earth's surface is based on the surface temperature, which varies widely around the globe as a function of space and time.

The heat that is transferred by convection flows through the troposphere at a rate determined by the temperature gradient of the atmosphere, or lapse rate, which is approximately 6.5°C/km in the troposphere. The heat that is transferred from the Earth's surface by radiative emission of LWIR and absorbed by CO_2 and H_2O molecules is transferred through molecular collision to nearby air molecules, mainly N_2 and O_2. Some of that heat is lost by convection to the upper atmosphere; some of that heat is radiated as LWIR upwards into space; and the balance of that heat energy is radiated as LWIR downward to the Earth's surface. *The amount of that heat energy is de minimis and has no effect on the temperature of the Earth's surface. There is no scientific basis for the claim that increased concentration of CO_2 in the troposphere causes global warming.*

The UAH Global Temperature Database of the Troposphere

Dr. John R. Christy is the Distinguished Professor of Atmospheric Science and Director of the Earth System Science Center at the University of Alabama in Huntsville where he began studying global climate issues in 1987. Since November 2000, he has been Alabama's State Climatologist. [24] In 1989, Dr. Roy W. Spencer (then a NASA/Marshall scientist, and now a Principal Research Scientist at UAH) [25] and Christy developed a global temperature dataset from microwave data observed from NOAA satellites, beginning in 1979. [26] For this achievement, the Spencer-Christy team was awarded NASA's Medal for Exceptional Scientific Achievement in 1991. In 1996, they were selected to receive a Special Award by the American Meteorological Society *"for developing a global, precise record of Earth's temperature from operational polar-orbiting satellites, fundamentally advancing our ability to monitor climate."* In January 2002, Christy was inducted as a fellow of the American Meteorological Society." [24]

Figure 13 depicts the UAH monthly global lower troposphere temperature anomaly readings for the period Dec. 1, 1978, to July 2020. [26]

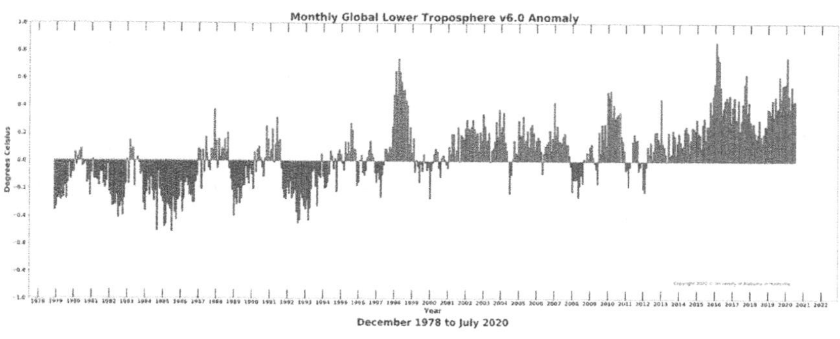

Figure 13. UAH monthly global lower troposphere temperature anomaly readings for the period December 1978 to July 2020. Graph used courtesy of University of Alabama Huntsville. [26]

The figure above also reveals a global lower troposphere temperature anomaly of +0.134°C per decade since Dec. 1,1978. [26] It should be noted that from 1979 to 1998, there was a global cooling period. The period 1998 to 2013 depicts a slowdown in the growth of global warming known as the "global warming hiatus or pause." However, over the 40-year period, the aggregate decrease in the temperature anomalies is roughly equal to the increase. Therefore, what can one reasonably conclude from these readings? The annual global troposphere temperature anomaly measurements are highly variable; and 40 years is too short a time scale to draw a meaningful conclusion as to longer term trends. Finally, it should be noted that based on the temperature database that exists to-date, a temperature anomaly of +0.134°C per decade, or +0.0134°C per year, is not statistically significant and arguably within the measurement margin of error.

CHAPTER TEN

THE THERMODYNAMIC INTERACTIONS WITH THE EARTH'S OCEANS

THE OCEANS ON EARTH cover 71% of the Earth's surface and contain 97% of the Earth's water. Less than 1% of the Earth's water is fresh water, and 69% of that amount is contained in glaciers and ice caps. The oceans contain 99% of the habitable space on the planet. Ninety-four percent of life on Earth is aquatic.[1] That makes man a minority in the animal kingdom!

So, just how large are the world's oceans? The total surface area of Earth is 510,082,000 sq. km (square kilometers). The oceans cover ~71%. Therefore, the total surface area is (0.71) (5.10×10^8) sq. km. = 3.621 x 108 sq. km. or 362 million square kilometers. What is the average depth of the world's oceans? The National Oceanic and Atmospheric Administration (NOAA) has calculated the average depth of the world's oceans to be 12,100 ft, or 3,688 meters (m), or 3.688 kilometers. [2] Therefore, the world's oceans have a water volume of:
(3.61×10^8) sq. km. x (3.688) km. = 13.31×10^8 cu. km., or 1.331×10^9 cu. km.

How many gallons of water are in the Earth's oceans? A cubic kilometer contains 2.642×10^{11} U.S. liquid gallons. Therefore, the world's oceans contain:

(1.331×10^9) cu. Km. (2.642×10^{11}) gal/cu. km. = 3.516×10^{20} gallons of water.

In numerical terms, that number is 351,600,000,000,000,000,000 gallons!

The scale of the world's oceans presents particular challenges in terms of the ability to acquire and analyze data for scientific purposes to predict changes in ocean parameters, such as temperature, pressure, and salinity. In addition, the movement of ocean waters around the globe results in energy and mass transfers, which are very difficult to model using computers with any degree of accuracy. Having said that, let's examine some of the history of man's efforts to acquire data about the world's oceans.

How do Scientists Measure the Mean Temperature of the World's Oceans Combined?

Accurate historical datasets of the average sea surface temperature (SST) of the Earth do not exist prior to 1978, when the National Oceanic and Atmospheric Administration (NOAA) launched satellites to orbit the Earth to measure the temperature of the troposphere and collect data on sea surface temperature ("SST").[3]

For the past 150 years or so, the SST measurements have been taken by ships. [3] SST was one of the first oceanographic variables to be measured. Benjamin Franklin suspended a mercury thermometer from a ship while traveling between the United States and Europe in his survey of the Gulf Stream in the late eighteenth century. SST was later measured by dipping a thermometer into a canvas or metal bucket of water that was manually drawn from the sea surface. Measurements of SST have had inconsistencies over the last 130 years due to the way they were taken. In the nineteenth century, measurements were taken in a bucket off a ship. However, there was a slight variation in temperature because of the differences in buckets. Samples were collected in either a wooden or an uninsulated canvas bucket, but the canvas bucket cooled quicker than the wooden bucket. [3]

The sudden change in temperature between 1940 and 1941 was the result of an undocumented change in procedure. The samples were taken near the engine intake because it was too dangerous to use lights to take measurements over the side of the ship at night. [4] The first automated technique for determining SST was accomplished by measuring the temperature of water in the intake port of large ships, which was underway by 1963. These observations have a warm bias of around 0.6°C (1°F) due to the heat of the engine room. This bias has led to changes in the perception of global warming since 2000 when the historical readings were "adjusted" downward in an effort to correct the bias. [5]

Many different drifting buoys exist around the world that vary in design, and the location of reliable temperature sensors varies. Fixed weather buoys measure the water temperature at a depth of three meters (9.8 ft). These measurements are beamed to satellites for immediate automated data distribution. A large network of coastal buoys in U.S. waters is maintained by the National Data Buoy Center (NDBC). Between 1985 and 1994, an extensive array of moored and drifting buoys was deployed across the equatorial Pacific Ocean designed to help monitor and predict the El Niño phenomenon.[4]

The temperature of the surface waters in the oceans varies mainly with latitude. The polar seas (high latitude) can be as cold as -2°C (28.4°F), while the Persian Gulf (low latitude) can be as warm as 36°C (96.8°F).

Ocean water, with an average salinity of 35 psu, freezes at -1.94°C (28.5°F). That means at high latitudes sea ice can form. The "average" temperature of the ocean surface waters as measured by these various methods is reported to be about 17°C (62.6°F).[6] However, it should be obvious from the temperature ranges of the ocean waters at various latitudes that *an average temperature of the oceans, like that of the Earth, is an abstraction and has no value in the physical world for analytic purposes.*

Ninety percent of the volume of the world's oceans is below the thermocline (200m or 660 ft.), where the temperature is between 0 - 3°C, or 32 - 37.5°F. [6] The Argo Float Program, started in the year 2000, is a collaborative partnership of more than 30 nations from all continents. It deploys over 4,000 free drifting floats, which are programmed to sink to a depth of 2,000 meters for ten days. Then, on the way back to the surface, the Argo floats measure temperature, currents, and salinity of the water.[7] For the first time, the system of Argo floats will provide data that will allow climate scientists to accurately measure some important variables that might influence thinking regarding the state of the world's oceans and Earth's climate. Historically, the lack of data has led to some unfounded statements about a purported linkage between AGW and the world's oceans.

I have listed here a few examples of the hyperbole that tends to accompany claims about man-made global warming and the oceans:

"National Geographic" ~ April 27, 2010:
"Global warming caused by human activities that emit heat-trapping carbon dioxide has raised the average global temperature by about 1°F (0.6°C) over the past century. In the oceans, this change has only been

about 0.18°F (0.1°C). ***This warming has occurred from the surface to a depth of about 2,300 feet (700 meters), where most marine life thrives*** *(emphasis added)." [8]*

The fact of the matter is that there is no reliable temperature dataset for the world's oceans that covers the period 1910 to 2010. Not only that, but "National Geographic" apparently does not agree with NASA, NOAA or the climate scientists who **know** that the oceans have "trapped" 90% of the "global warming caused by human activities that emit heat-trapping carbon dioxide (which) has raised the average global temperature by about 1°F (0.6°C) over the past century." (See below)

This is an excerpt from the Executive Summary of the report of Working Group 1 of the UN IPCC, in the 5th Assessment Report published in 2013, entitled, *"Observations: Ocean."* [9] The bracketed numbers refer to either the corresponding sections of the report or ranges of uncertainty in the estimates:

"It is virtually certain *(emphasis added) that the upper ocean (above 700 m) has warmed from 1971 to 2010, and* ***likely that it has warmed from the 1870s to 1971*** *(emphasis added). Confidence in the assessment for the time period since 1971 is high based on increased data coverage after this date and on a high level of agreement among independent observations of subsurface temperature [3.2], sea surface temperature [2.4.2], and sea level rise, which is known to include a substantial component due to thermal expansion [3.7, Chapter 13]. There is less certainty in changes prior to 1971 because of* ***relatively sparse sampling in earlier time periods.*** *(emphasis added) The strongest warming is found near the sea surface 0.11°C [0.09 to 0.13] per decade in* ***the upper 75 m between 1971 and 2010*** *(emphasis added), decreasing to about* ***0.015°C per***

***decade at 700 m** (emphasis added)."* Due to the imprecise nature of the reporting by various sources on supposed AGW, it is often difficult to compare the reported data. For example, in the above IPCC report, the weighted average increase in temperature per decade, for the period 1971 to 2010, is 0.024°C from the surface down to a depth of 700 m. The "National Geographic" reported an increase of 0.01°C per decade, for the same depths, for the period 1910-2010. So, which report is one to believe?

The following excerpt is from an article in 2014 on NOAA's website: *"Climate scientists **know that the ocean has stored about 90 percent of the Earth's extra heat trapped by man-made greenhouse gases since 1955** (emphasis added). Using data from a global array of Argo Floats – aquatic robots that measure temperature and salinity at different ocean depths – **scientists have since learned that as air temperatures have increased in the atmosphere, the surface ocean has been taking up an increased amount of heat."** (emphasis added)* [10]

As was previously discussed, the Argo float program did not begin until the year 2000. Therefore, no accurate data exists on ocean temperatures prior to that time, including the period 1955 - 2000. The reader is not told the quantity of "extra heat" that was "trapped" since 1955. **The long-term averages of the surface air temperatures above the ocean approximate 2°C below the sea surface temperature. Therefore, there can be no heat transferred from the atmosphere to the ocean since such an event would violate the Second Law of Thermodynamics.**

This is from a NASA news release in July 2015:
*"A new NASA study of ocean temperature measurements shows that **in***

recent years, extra heat from greenhouse gases has been trapped *(emphasis added) in the waters of the Pacific and Indian oceans. Researchers say this* ***shifting pattern of ocean heat accounts for the slowdown in the global surface temperature trend observed during the past decade*** *(emphasis added)."* [11]

The last sentence appears to be an effort to explain the global cooling trend depicted in the global temperature satellite datasets during the period 1998 to 2013, known as the "global warming hiatus." It implies that the "extra heat from greenhouse gases" (assumedly caused by man-made activities) that would have increased the "global surface temperature" was transferred into the Pacific and Indian oceans and was "trapped" there. As stated earlier, the long-term surface air temperature above the oceans is lower than the sea surface temperature. **The atmosphere cannot transfer heat to the oceans in violation of the Second Law of Thermodynamics.**

I find it interesting that so many of these articles and many of the proponents of the global warming hypothesis, including some climate scientists, use the term "trapped heat" in what are supposedly scholarly essays. The Cambridge English dictionary defines trap as "if something is trapped, that person or thing is unable to move or escape from a place or situation." The use of the word trapped in this context implies that the heat is unable to move and is "unable to escape" the ocean or a CO_2 molecule.

Heat cannot be trapped! Heat is a form of energy, and it is constantly on the move. A corollary of the Zeroth Law of Thermodynamics is that if a body is not in thermal equilibrium with an adjacent body and they share a boundary that is permeable to heat, energy will spontaneously flow from the warmer body to the cooler body. If any of the world's oceans were to gain heat by interactions with a warmer body of water or by absorbing more solar energy, that heat gained would be transferred to a cooler thermodynamic body over time (land mass, water, or atmosphere) until the Earth's biosphere achieved thermal equilibrium, which as a practical matter, cannot occur. As long as a thermal gradient (difference in temperature) exists in the Earth's atmosphere, land mass or in the world's oceans, heat will flow from a warmer body to a cooler body. **That is the law!**

The fallacy in the NASA statement above is that heat can flow from the atmosphere into the oceans; or that LWIR from greenhouse gases can penetrate the ocean's surface and heat the waters below. It is not physically possible for either of these events to occur! As will be demonstrated later in this chapter, downwelling LWIR from the Earth's atmosphere cannot penetrate the surface of the ocean below 100 microns (about the width of the average human hair). It can have **no** effect on the ocean's temperature.

On August 10, 2017, NOAA released this statement about 2016 mean ocean temperatures:
"Global upper-ocean heat content neared record high. ***Heat in the uppermost layer of the ocean, the top 2,300 feet (700 meters), saw a slight drop*** *compared to the record high set in 2015 (emphasis added).* ***The findings are consistent with a continuing trend of warming oceans*** *(emphasis added)."* [12]

Perhaps like me, you are confused by the conclusion that NOAA stated based on its introductory sentence. How could NOAA conclude that a "slight drop" in the temperature of the top 2,300 feet of the ocean depths, where the most heat is believed to be "stored," be consistent with a "continuing trend of warming of the oceans?" The misleading nature of this statement is typical of the statements made by the proponents of AGW when the facts contradict the hypothesis.

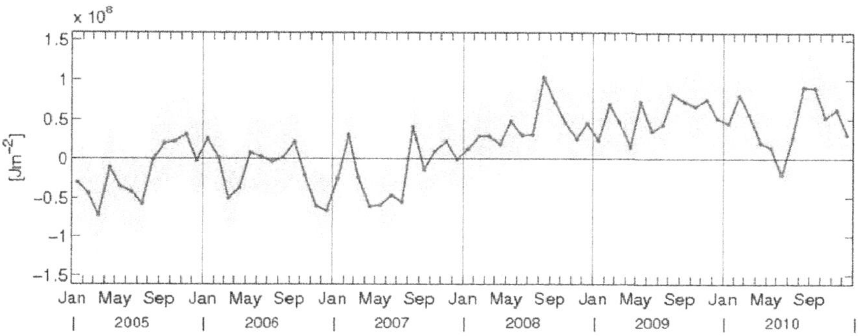

Figure 14. "Ocean Heat Content for 10-1500m Depth Based on Argo: 2005 – 2010." Content used courtesy of NOAA. [13]

As one can see from Figure 14 above, from 2005 to 2012 the ocean heat content (OHC), as measured by the Argo float program, indicated very little change in overall OHC (+0.54 $W{\cdot}m^{-2}$, or 0.067 $W{\cdot}m^{-2}$/yr.), in direct contradiction of the published statements made by NASA and NOAA in 2014, 2015 and 2017. [13]

On February 13, 2020, the NOAA website, climate.gov, published this statement:

"Averaged over the full depth of the ocean, the 1993 – 2018 heat-gain rates are 0.57 – 0.81 watts per square meter." [14] It should be noted that a value of 0.81 $W{\cdot}m^{-2}$ of heat gain over a 25-year period (0.032 $W{\cdot}m^{-2}$/yr. on average) is de minimis in comparison with the value of maximum solar irradiance on a clear day (1000 $W{\cdot}m^{-2}$), and arguably, well within the

measurement margin of error. Although NOAA does not report the heat gain as a function of specific time periods, the average heat gain rate of 0.032 $W{\cdot}m^{-2}$/yr. is less than one-half of that reported by the Argo floats at 0.067 $W{\cdot}m^{-2}$/yr. In either case, the heat gain is *de minimis.*

Notwithstanding the difficulty in accurately computing such a small value over such a large volume and such a long timeframe, when compared to the maximum value of solar irradiance on a clear day (~1000 $W{\cdot}m^{-2}$), such a value seems inconsequential. **If** one were to accept this calculation to be correct, how could it suggest warming of any significance?

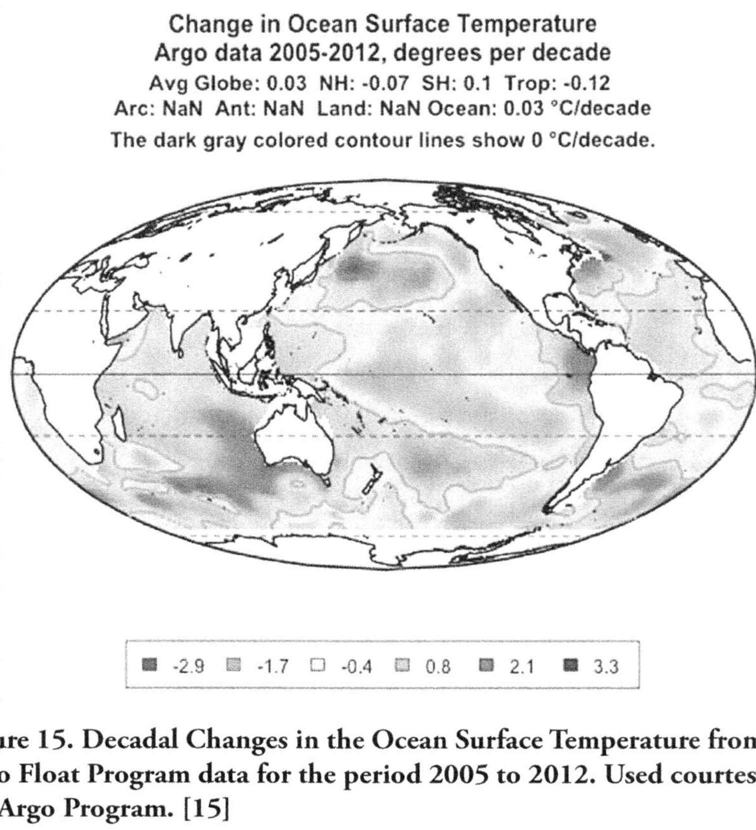

Figure 15. Decadal Changes in the Ocean Surface Temperature from Argo Float Program data for the period 2005 to 2012. Used courtesy of the Argo Program. [15]

Figure 15 above depicts the decadal changes in the average ocean sea surface temperature (SST) for the period 2005-2012, as measured by the Argo Float program. The global average SST change per decade is 0.03°C. The map depicts large areas of the oceans' SST that have cooled by 1.7°C or 0.4°C and some have warmed as much as 2°C. [15] As previously discussed, average temperatures have no meaning in thermodynamic analyses; the uneven distribution and magnitude of temperature change suggests that more data is required to understand the causes. However, the stated overall increase is *de minimus* and arguably within the measurement margin of error when averaged over such a large volume. In addition, an eight-year period is too short a time frame to infer any trends.

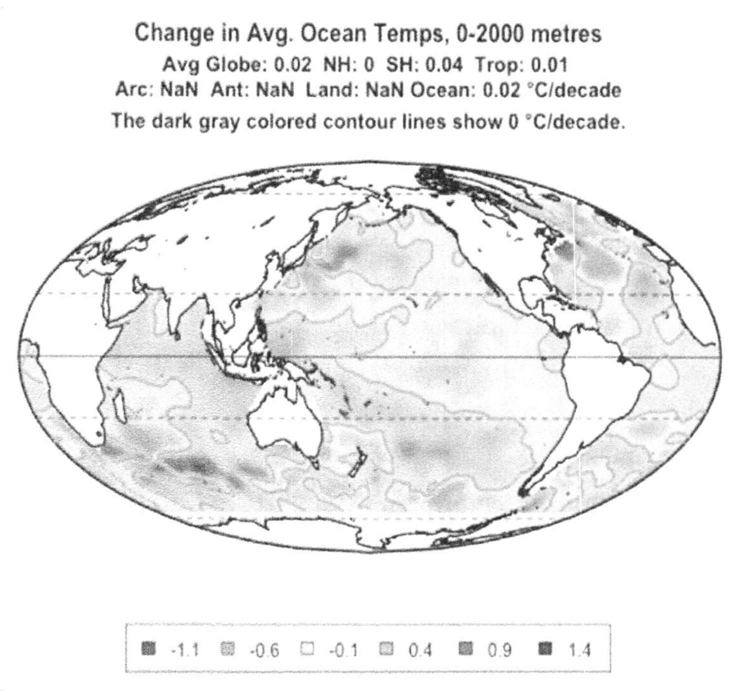

Figure 16. "Decadal change in average ocean temperatures 0 — 2000 meters based on Argo Float Program Data for the period 2005-2012." Used courtesy of the Argo Program. [16]

Figure 16 illustrates the decadal change in average ocean temperatures from a depth of 0 – 2,000 meters based on the data acquired from the Argo floats for the period 2005-2012. The overall change in the average temperature across these depths is +0.02°C per decade. [16] If the change in the ocean's average temperature at varying depths has been negligible over the last eight years, **the claim that the oceans have "trapped heat" that was generated because of man-made activities, presumably because of the increase in CO_2 concentration in the Earth's atmosphere, is clearly false.**

As was discussed in Chapter 9, the increase in the downward net heat flux from the lower troposphere because of the increase in the CO_2 concentration in the Earth's atmosphere from 280 ppm to 400 ppm over the last 60 years is calculated to be approximately 2 $W \cdot m^{-2}$. This net heat flux is the result of LWIR emitted by the blackbody continuum of CO_2 and H_2O molecules, primarily H_2O, located within the first kilometer of the atmosphere. Let's examine the underlying science that governs the thermodynamic interaction of the lower troposphere and the ocean to understand why, aside from the fact that the value is *de minimis,* that the LWIR emissions from the troposphere have had no effect on the ocean's temperatures.

Water is almost completely opaque to LWIR radiation. LWIR from the Earth's atmosphere cannot penetrate the surface of the world's oceans below a depth of 100 microns. The LWIR absorption/emission interaction volume is at most 10 cm^3. Based on the Kirchhoff Exchange Law, an increase in downward net LWIR flux at the ocean's surface decreases the ocean LWIR cooling flux by a corresponding amount.[15] However, the ocean responds by rapidly increasing the surface evaporation rate to counteract the heat exchange increase. Therefore, there is no net heat gain from the *de minimis* LWIR heat flux absorbed by the world's oceans. The long-term averages of the surface air temperatures above the ocean approximate 2°C below the sea surface temperature. [15] Therefore, in concert with the Second Law of Thermodynamics, the atmosphere cannot heat the ocean. **Finally, it is impossible for an increase in downward atmospheric LWIR flux of 2 W·m^{-2} to heat the ocean.** The slight increase in the heat flux is converted by the ocean surface into an insignificant change in evaporation rate. Solar irradiance can penetrate the surface of the ocean to a depth of 100 m. [15] Any heat gain in the world's oceans is the result of solar irradiance, not LWIR.

In conclusion, the temperature datasets from the Argo Float Program indicate that negligible (0.02°C/decade) temperature change has occurred in the first two kilometers of the Earth's oceans for the period 2005 to 2012. Furthermore, scientific analysis using spectroscopic data for the transmissivity of LWIR in seawater demonstrates that LWIR cannot penetrate the ocean's surface below 100-micron depth [15]. Finally, the laws and principles of thermodynamics demonstrate that the negligible heat flux (2 W·m^{-2}) generated from an increase in concentration of 120 ppm CO_2 in the Earth's atmosphere is offset by a commensurate increase in evaporative cooling. **Therefore, any increase in CO_2 concentration in the Earth's atmosphere due to anthropogenic activities has had no effect on the temperatures of the world's oceans.**

CHAPTER ELEVEN

EMPIRICAL EVIDENCE OF GLOBAL WARMING

IN ALMOST EVERY CONVERSATION concerning the subject of man-made global warming with a proponent of the AGW hypothesis, that person will cite empirical evidence to support the argument: arctic temperatures are at record highs; arctic ice is melting at a record rate; there has been a decrease in Antarctic shelf ice; there are a record number of wildfires around the globe; and the world's weather is setting new heat records at various locales. Usually, the empirical evidence is one-sided and does not include examples of weather extremes that refute the hypothesis. It is important to understand the difference between weather and climate. Weather reflects the short-term conditions of the atmosphere and is affected by temperature, pressure, humidity, cloudiness, wind, precipitation, flooding, ice storms, and other local phenomena. Climate is the average daily weather for an extended period of time at a certain location. Climate is what you expect; weather is what you get. [1]

On July 21, 1983, the lowest recorded temperature ever at a land-based weather station on Earth was recorded at the Soviet Vostok Station in Antarctica, -89.2°C (-128.6°F). [2] Satellite readings at approximately 100 sites in the Antarctic have observed minimum surface temperatures of -98°C during the winters of 2004 to 2016. Comparisons of surface snow temperatures with near-surface air temperatures at nearby weather stations indicate that -98°C surfaces imply -94 ± 4°C air temperatures at 2 meters above the surface. [3]

However, the story is not the same in the Arctic. According to a study published in Science Advances, in December 2019, the Arctic has warmed by 2°C to 3°C since the late 19th century. [4] Over the past decade, the Arctic has warmed by 0.75°C, far outpacing the global average, while Antarctic temperatures have remained comparatively stable. Finally, this according to a NASA press release in September 2019 [5]: *"Arctic Sea ice reaches its minimum each September. September Arctic sea ice is now declining at a rate of 12.85 percent per decade, relative to the 1981 to 2010 average."*

Verkhoyansk is a town in Russia, located on the Yana River in the Arctic Circle at approximately 67.5°N. According to the World Meteorological Organization records, the highest temperature ever recorded in Verkhoyansk was recorded at 37.3°C on July 25, 1988. Fort Yukon, in Alaska, located at 66.5°N, recorded the first ever 37.7°C temperature in 1915, long before the recent increase in CO_2 concentration in the Earth's atmosphere could have had any effect on global temperatures. [6] On July 25, 1988, Fort Yukon recorded a high temperature of 23.2°C. So, what is one to conclude about changes in climate from these empirical events regarding temperature readings? Did a global cooling period from 1915 to 1988 cause the high temperature in Fort Yukon to drop from 37.7°C in 1915 to 23.2°C in 1988? *The fact is that record high and low temperature readings will occur over time in places around the Earth and those temperature readings have no correlation to climate change.*

Foehn Winds

Foehn winds are a type of dry, warm, down-slope wind that occurs in the lee (downwind side) of a mountain range. Winds of this type occur when moist, cool air ascends a mountain slope. As the air nears the top, it cools, leading to condensation and latent heat release that reduces the temperature of the air, which results in precipitation that removes the condensed water. The descent on the lee side is dry, which increases the (pressure related) warming, leading to higher lee-side temperatures. This warming can be spectacular (e.g., 25°C in an hour; Richner and Hächler 2013) [7] and is typically accompanied by a decrease in humidity with accelerated downslope winds with speeds up to 100 k per hour (60 mph). [8] In the Arctic, warm, dry foehn winds can cause melting at the edges of glaciers and have been recorded over the White Glacier of Axel Heiberg Island. [9] According to research published on April 11, 2019, in *Geophysical Research Letters,* foehn winds in the Antarctic - which are at their strongest outside of the summer season - have caused significant late-season melting on the Larsen C ice shelf each year since 2015. [10] In conclusion, foehn winds could have a significant, transitory, local effect on polar and Antarctic ice formations. In addition, foehn winds around the world, such as the Santa Ana winds in California, amplify the effects of wildfires. There is no proven linkage between the presence of foehn winds and man-made climate change.

Arctic Ice and the Atlantic Multi-decadal Oscillations

The average ocean heat content, which cannot be *measured,* only calculated, and like the average land surface temperature, has no real meaning in scientific analysis, is reported to have only gained around .81 $W{\cdot}m^{-2}$ over the last 25 years. When compared to the daily total solar irradiance received by the world's oceans, is a *de minimis* amount. As discussed earlier in this chapter, some of the world's oceans have reportedly warmed at 2°C per decade; some have cooled at 1.5°C per decade. There are complicated ocean currents at work throughout the world carrying heat from the equatorial regions towards both the north and south poles. No one knows the exact effect of those currents and gyres as they might impact polar ice extent. That effect can only be inferred by comparing data from other ocean phenomena such as the Atlantic Multi-decadal Oscillation (AMO), the summer surface air temperature (SAT) in the Arctic and measurements of the Arctic Sea ice extent (SIE). The difficulty in making exact comparisons lies in the fact that an accurate location-specific, measured, historical database for each of these variables does not exist before NOAA satellites began to circumnavigate the globe in 1979. Therefore, to develop the historical database depicted in various graphs before that time, scientists had to estimate that data using mathematical models.

The AMO is the average surface temperature of the North Atlantic basin from the equator to 60 N. *There appears to be a very strong correlation among AMO cycles, SAT values and the SIE.* [11]

An increase of SAT in the marine Arctic (the part of the Arctic covered with sea ice in winter) shows a good relationship with the reduction of SIE in summer. For instance, a strong correlation (a coefficient equal to –0.93) was found between the summer SAT in the marine Arctic and the satellite-derived, 1980 to 2014 September sea ice index, as depicted in Figure 17 below. [12]

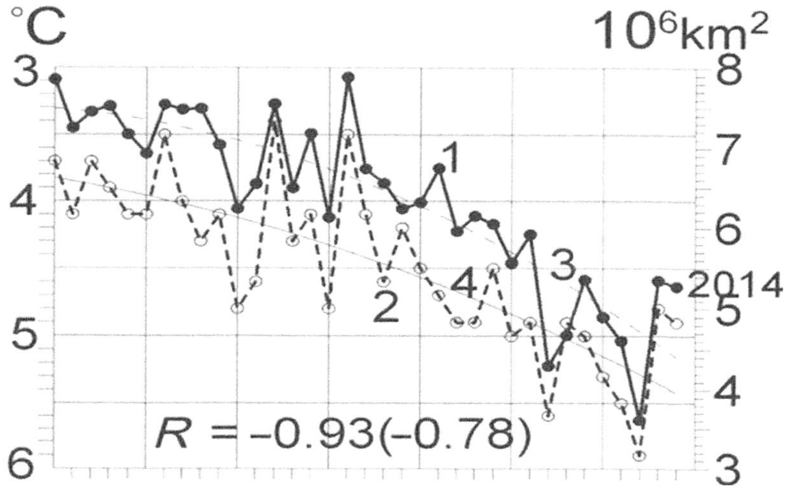

Figure 17: "A comparison of average Arctic SIE in September for the period 1980-2014 with the average summer surface air temperature." [12]

As seen in Figure 17, the plot of the average summer surface air temperature in the Arctic (solid black line) is inverted and depicts an increase in temperature during the period, from 3°C in 1980 to 5.6°C in 2012. The SIE (dashed line) follows that trend very closely, depicting a decrease in Arctic SIE from around 6.8 million square kilometers in 1980 to around 3.2 million square kilometers in 2012, with an increase to 5.6 million square kilometers in 2014 as the summer surface air temperature decreased.[12]

The graph below in Figure 18 depicts a plot of the Atlantic Multidecadal Oscillation for the period 1856 to 2020. The temperature increase is 0.028°C per decade, or 0.46°C, over the 164-year period.[12]

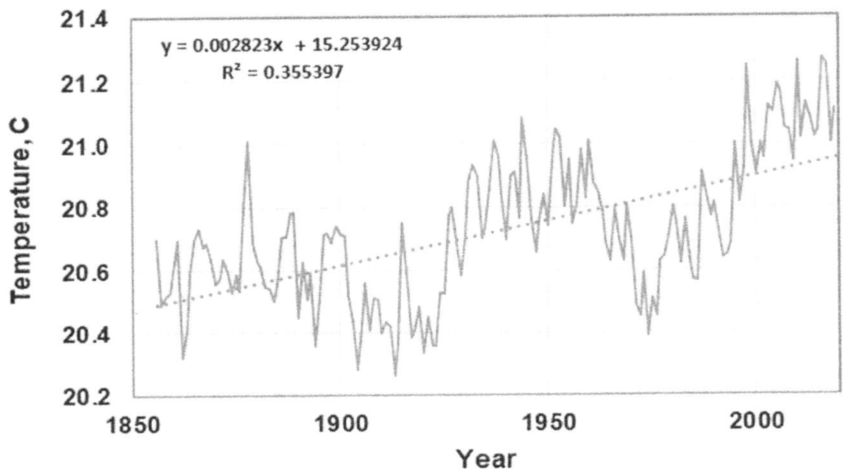

Figure 18: "The Atlantic Multi-Decadal Oscillation for the period 1856-2020. Courtesy of NOAA. [12]

Based on the data plots in Figure 18, the AMO has experienced three cycles since the 1850s of about 60 to 80 years in length. The correlation between the AMO cycles and SIE in the Arctic has been further corroborated with anecdotal evidence in historical commercial and military shipping logs that confirmed a reduction in the SIE during the periods when the AMO entered a warming phase. [12]

There is also a strong correlation between historical Meteorological Surface Air Temperature Station (MSATS) readings as measured by the UK Met Office HadCRUT4 and the AMO as depicted in Figure 19 below.

As can be seen in Figure 19, a very close correlation between the AMO index and the HadCRUT4 temperature anomalies existed until around 1970. If one adjusts the HadCRUT4 data plot with an offset of -0.3°C for the period 1970 to 2020, the plots once again overlap very closely.

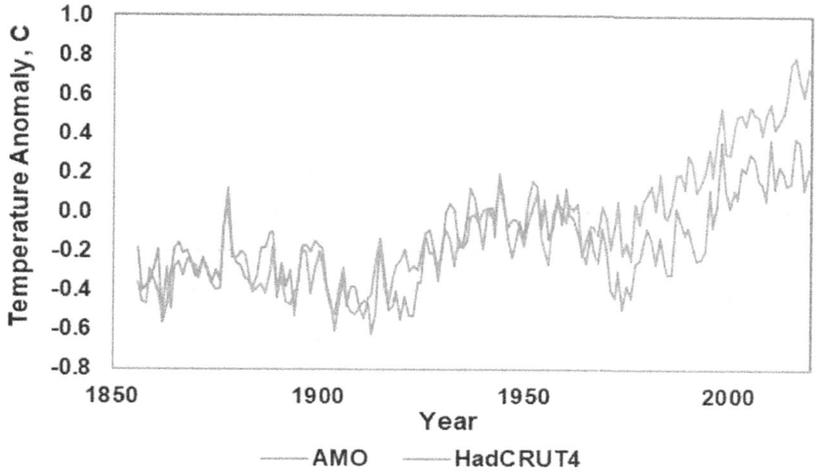

Figure 19: "A Comparison between the AMO Index and HadCRUT4 Surface Temperature Anomalies for the Period 1870-2020, Corrected for Changes in Station Data Adjustments" Courtesy of NOAA.

[12] The prospective reasons for the 0.3°C increase in the temperature anomalies depicted in the HadCRUT4 data for this period include major changes in the number of weather stations used in the database, a change from manual to electronic data recording, the advent of urban heat island effects and *the use of homogenization techniques to adjust the data.* [12]

So, what can one reasonably conclude from the above analysis? Cyclical changes in the sea surface temperature of the North Atlantic Basin cause cyclical changes in the extent of polar ice. When the AMO rises, sea ice melts. When the AMO falls, sea ice grows. In addition, there is a strong correlation between the AMO temperature anomalies and the MSAT temperature anomalies. Now, we must revert to first principles of science to deduce what causes that correlation. Does the atmosphere heat the ocean or vice versa? As we learned in Chapter 10, the long-term averages of the surface air temperature above the world's oceans approximate 2°C below the sea surface temperature. Therefore, in accordance with the Second Law of Thermodynamics, the atmosphere cannot heat the ocean. As a result, we must conclude that the oceans are heating the atmosphere on some periodic basis. Given that the world's oceans are only heated by the Sun (and not LWIR photons emitted from the atmosphere), then one must further logically deduce that changes in solar irradiance, in conjunction with the thermal capacity of the oceans (the ability of the oceans to absorb or give up heat energy for a period of time before a unit change in temperature occurs), are producing the AMO cycles. Finally, what is causing the periodic change in solar irradiance? Is it the sunspot cycle? Is it the Milankovitch cycle? Could other unknown factors affect the total value of solar irradiance? The answer to each of the preceding questions is yes. Are man-made activities affecting the value of total solar irradiance or the temperature of the world's oceans? No.

The Antarctic Ice Sheet

As it does in the Arctic, the surface of the ocean around Antarctica freezes over in the winter and melts each summer. Antarctic sea ice usually reaches its annual maximum extent in mid to late September and reaches its annual minimum in late February or early March. The 2020 minimum extent (February 20 – 21, 2020) was below the 1981 to 2010 climatological average, but well above the record low recorded in 2017. [12]

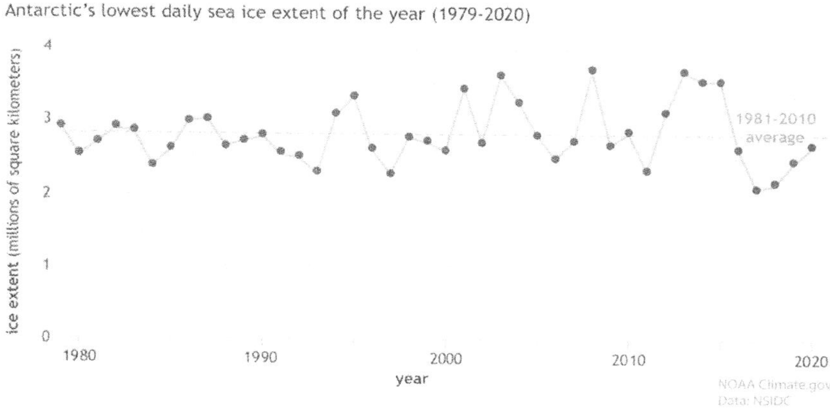

Figure 20: Antarctic Ice Sheet Extent for the period 1979-2020. Courtesy of NOAA. [12]

As can be seen in Figure 20, the Antarctic ice sheet extent in 2020 was slightly greater than the 1980 value. In fact, the trend for Antarctic ice sheet extent growth is positive for the period 1995 to 2020, notwithstanding a decrease in 2016 and 2017. Almost all of the UN IPCC CMIP5 models predict a decrease in Antarctic Sea ice based on the anthropogenic global warming hypothesis. So, what is causing the increase in Antarctic Sea ice? Climate scientists have no explanation for the deviation from the model predictions.

The fact is that there is a lot we don't know about the factors that affect the climate on Earth, especially in the polar regions. While the NOAA satellites that were launched in 1979 were an important first step in gathering data about the troposphere, 40 years is just the blink of an eye in terms of climate history. The Argo float program was initiated in 2000 and is a good start in monitoring the oceans' temperature, salinity (which affects the density and therefore movement of ocean water) and currents. However, the 4,000-float sampling density is still low (one float per 90,500 square kilometers) and the elapsed time too short (20 years) to develop a reliable database.

The atmosphere and the world's oceans are large and the current scientific data available on these thermodynamic bodies is limited. Absent additional hard data, it is not possible to explain why these empirical events occur with any reasonable degree of certainty.

THE FACTS BEHIND THE CLAIMS OF MAN-MADE GLOBAL WARMING

"97% of the world's scientists agree that man has caused global warming."

It seems that the "citizen science project" conducted by volunteers, led by John Cook, and published on May 15, 2015, was the seminal work that led to the claim that "97% of the world's scientists agree that man has caused global warming." While not a peer-reviewed project, the authors structured the report in such a way as to imply that it followed the basic outline for a research project. However, *the authors of approximately 66% of the 11,944 papers reviewed expressed no opinion on global warming.* The 97% was derived from the other 34% that did express an opinion. [1]

Why would 66.4% of the papers that contained the phrase "global climate change" or "global warming," (which were published during the ascendancy of the man-made global warming issue) not express an opinion on what could arguably be called the most important subject in the history of climate science?
In a second survey, only 14% of the authors responded and it was reported that 97% of those 'endorsed the consensus.'

How could it be possible that any unbiased reader who carefully reviewed the methodology, results and conclusions reported in this "citizen science project" accept the conclusion from the study that 97% of the world's scientists agree that man has caused global warming? How could any thinking person be duped by such illogic and manipulation of the mathematics in the report? How could anyone in the climate science community accept the authors' conclusions? How could any one of the investigators defend the logic employed in the study with a straight face?

I think that it is obvious that the investigators in the "citizen science project" had a bias towards a particular outcome. Why the claims of this study gained traction within the climate science community is a puzzle, unless it served to promote the AGW hypothesis and funding for climate research. Why would certain political leaders around the world be so quick to accept this claim?

Let's review the views of the world's science organizations set forth in Chapter 3 to see if we can gain insight into this issue.

The Views of the World's Universities and Academic Societies on AGW

I think that a review of this topic provides particular insight into the evolution of the claim that a consensus of thinking among the world scientific community exists regarding man-made global warming. Let's engage in a quick review of the positions of some of the world's science organizations and academies on the subject.

As you may recall, I began much of the basic research work on the AGW hypothesis in June 2017. At that time, I surveyed the websites of many of the world's universities and science academies to determine their position on AGW. In October 2019, I again surveyed the same organizations to see if their positions on AGW had changed. The results of that second survey surprised me. Apart from the Russians and the Chinese, several organizations had "evolved" their thinking in that two-year period to accept the AGW hypothesis. Most cited either the "research" work of the UN IPCC or anecdotal evidence such as "forests are drying out, glaciers are melting, hurricanes are raging in the Caribbean" as the basis for their changed view.

So, what is one to make of the quick reversal of opinion by the British and Germans? Why do the U.S. Academy of Sciences and NASA continue to endorse the AGW hypothesis? Why the insistence on the part of proponents of the AGW hypothesis to squelch the debate by declaring that "the science is settled?" Why the move towards "group think" led by the UN IPCC? **It's all about the money.**

Expenditures on Climate Research exceed $1 Trillion

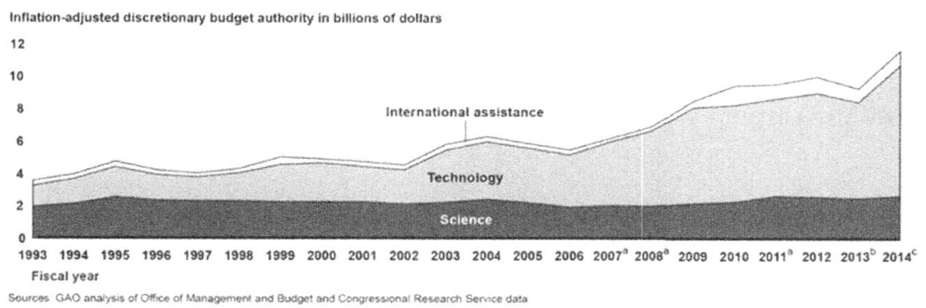

Figure 21. U.S. Government Funding for Climate Change Research for the Period 1993 – 2014.[2] Courtesy of United States Government Accountability Office.

Since the inception of the UN IPCC in 1988, the funding for climate related research has surged and the U.S. has led the way, as is outlined in Figure 21 prepared by the U.S. General Accounting Office. [2] There are no reliable estimates on the total amount of money that has been spent worldwide on climate change in the last 20 years or so. However, various reports put the figure at well over $1 trillion. In 2013, the "Climate Policy Initiative" issued a study which found that "global investment in climate change" reached $359 billion that year. [3]

According to a recent report by the U.S. Government Accountability Office, "Federal funding for climate change research, technology, international assistance, and adaptation has increased from $2.4 billion in 1993 to $11.6 billion in 2014, with an additional $26.1 billion for climate change programs and activities provided by the American Recovery and Reinvestment Act in 2009." [2] Both Presidents Bill Clinton and Barack Obama made climate change a central part of their presidency. President Obama made the issue of climate change his legacy. *"Climate change,"* Mr. Obama often said, *"is the greatest long-term threat facing the world."* [4] It should be noted that the U.S. Government spent $5.1 billion per year on average over the period 2012 to 2017 on cancer research, while more than twice that amount per year was spent on researching climate change. [5]

The fact of the matter is that every scientific organization and individual climate scientist who wants to do climate research has to depend on outside funding in some form or fashion to pay for that research. While the energy industry funds both sides of the climate debate, the government/foundation monies go only toward research that advances the warming regulatory agenda. It is no wonder that NASA, the U.S. Academy of Sciences, the Helmholtz Association, the British Royal Society, the French Academy of Sciences, and other organizations that depend on government financial support "endorse the consensus," especially when the heads of government promote the AGW hypothesis.

Politicians Support the AGW Hypothesis

It is not surprising that the Helmholtz Association would change their view so quickly in the course of two years.[6] *"Global warming is real. It is threatening. We must do everything humanly possible to overcome this challenge for humanity. This is still possible,"* said the German Chancellor in a speech on December 31, 2019. [6-8] No surprise that the Helmholtz Association quickly adopted the UN IPCC position between 2017 and 2019.

Prior to attending a climate summit in Paris in 2018, British Prime Minister Theresa May stated, *"There is a clear moral imperative for developed economies like the UK to help those around the world who stand to lose most from the consequences of man-made climate change."*[9] Therefore, it should be no surprise that the British Royal Society abandoned the research efforts to "discern the human influence on climate" and the view that scientists must "consider many natural variations that affect temperature, precipitation, and other aspects of climate from local to global scale, on timescales from days to decades and longer."

I believe that Prime Minister May's remarks reveal several important aspects of the thinking of politicians in Western Europe (and perhaps the US) as regards the issue of man-made climate change. Prior to Brexit, there was a collectivism in the thinking of the political leaders in the EU. President Macron was a strong proponent of internationalism and the man-made climate change hypothesis. He strongly encouraged other political leaders in Europe and the US to "get on board" with the

AGW hypothesis. In addition, I suspect that the Green Party influenced Macron's thinking as well as other European politicians. Also, I believe that there was/is a certain desire to be at the vanguard of socially acceptable thinking in supporting the AGW hypothesis. Finally, and perhaps of most concern, I don't think that many politicians understand the science involved in the AGW hypothesis, nor the social and economic ramifications of abandoning fossil fuels. Therefore, I think it is politically expedient to support the AGW hypothesis.

Finally, I offer this quote from Dr. Richard Lindzen concerning funding and peer pressure within the climate science community: *"Since this issue fully emerged in public almost thirty years ago - and was instantly incorporated into the catechism of political correctness - there has been a huge increase in government funding of the area, and that funding has been predicated on the premise of climate catastrophism.* ***By now, most of the people working in this area have entered it in response to this funding*** *(emphasis added)."* [10]

"Note that governments essentially have a monopoly over the funding. I would expect that the recipients of such funds would feel obligated to support the seriousness of the problem. ***Certainly, opposition would be a suicidal career move for a young academic*** *(emphasis added)."* [10]

Over the last several years, as I have researched the subject of AGW, I have found one aspect of the debate on the subject to be very revealing. Almost without exception, when a person of some stature in the scientific community states a position that opposes the AGW hypothesis, proponents of the hypothesis engage in *ad hominem* attacks against the opponent, rather than attacking the facts that purport to support the position. The intensity of those attacks tends to be directly related to the stature of the person within the scientific community. There are several websites which exist just to discredit the opponents of man-made global warming, and on occasion, their arguments.

The Scientific Method of Inquiry

For more than 400 years, from the 1600s to the 1900s, the Scientific Method of Inquiry served as the philosophical and empirical method of acquiring knowledge that has characterized the development of science. The Scientific Method emphasizes careful observation and a rigorous skepticism to what is observed, given that cognitive assumptions can distort how one interprets the observation. It involves formulating hypotheses (which can be falsified), and experimentation and measurement-based testing of the deductions drawn from the hypotheses. The final step of the scientific method requires the refinement or elimination of the hypothesis based on the experimental findings.

As discussed earlier, it would seem that a climate scientist who wished to investigate the prospect of whether man had caused global warming that would adversely affect life on Earth, might develop the following complex hypothesis to support such an investigation:

1. The temperature of the oceans, the Earth's land mass and relevant atmosphere has risen during the time since the advent of the Industrial Revolution, circa 1870 (which marks the beginning of man's alleged impact on global warming).
2. Man's activities are responsible for the global warming that has occurred beyond that caused by nature.
3. If global warming has occurred, the extent to which it has done so, or is reasonably projected to do so in the future, will adversely affect life on Earth.

It should be noted that each of the conjectures above are interdependent; that is, if any one of the conjectures is falsified, the complex hypothesis is falsified.

It is very clear that the climate science today that is employed in much of the debate promoting AGW has followed none of the dictates of the Scientific Method. First, it appears that no climate scientist or other researcher in the field has constructed a hypothesis or complex hypothesis such as above, as a result of careful observation of related phenomena while applying rigorous skepticism about what is observed.

Over a 40-year period, the NOAA satellites measured a temperature anomaly of only 0.13°C per decade in the troposphere. For the first 20 years of that 40-year period, the temperature anomalies indicated a *significant cooling* in the temperature of the troposphere. How could one who "applied rigorous skepticism to what is observed," possibly construe that such a limited dataset, with such a high degree of variability and conflicting trends in measured results over time, proves global warming? Further, the data from the Argo Float program indicates no appreciable heating of the world's oceans. In fact, certain parts of the world's oceans have experienced cooling.[11] Finally, the land surface temperature database, as flawed as it is, also depicts no statistically significant warming. None of the global temperature datasets depict appreciable global warming. The statements above serve to falsify the man-made global warming hypothesis.

What research, if any, has been done to try to eliminate natural causation from such a small variance in the temperature readings? In 2017, the British Royal Society, the oldest and most respected scientific organization in the world, recognized the need to study the potential effects of natural causation. However, the British Prime Minister pre-empted those efforts with a statement of support for AGW.

Finally, what scientific investigations have been done to determine what temperature rise of the Earth's oceans, land mass and relevant atmosphere would adversely affect life on Earth? Svante Arrhenius thought that a temperature increase of 4°C (7.2°F) would benefit mankind, "making the cooler climes more habitable and enhancing agricultural production around the globe." Based on the Paris Climate Accord statement, the proponents of the AGW hypothesis contend that an increase in the temperature of the oceans and Earth's relevant atmosphere greater than 1.5°C (2.7°F) will adversely affect life on Earth, *without any scientific studies to support such a claim.* It is implied in the current global warming manifesto that anyone who would even pose such a question would be considered immoral. Neither the second nor the third conjecture of the complex hypothesis above has even been considered in the discussion on man-made global warming.

The fact is that the Earth has gone through climate cycles in its recent past, and undoubtedly throughout its entire existence. The "Medieval Warm Period" is believed to have occurred during the period 950 A.D. through 1300 A.D. [12] Many anecdotal stories of the effects of the Little Ice Age include canals and rivers freezing across Europe, the destruction of farms and villages in the Swiss alps due to encroaching glaciers, the freezing of the Baltic Sea and famines throughout the world. [13] Theories as to what might have caused both climate changes include changes in solar irradiance, changes in volcanic activity and changes in the circulation of the world's oceans. *However, it can be stated with unequivocal certainty that man did not cause the climate change during these time periods; both occurred before the start of the Industrial Revolution.*

The United Nations Intergovernmental Panel on Climate Change

When it was formed in 1988, the stated mission of the UN IPCC was "to provide the world with a ***clear scientific view on the current state of knowledge*** *in climate change and its* ***potential*** *environmental and socio-economic impacts (emphasis added)."* The IPCC failed to pursue its mission. There were no reports on the state of the knowledge in the science of climate change. The AGW hypothesis was accepted as a fait accompli. There were no reports about the varying scientific views regarding the subject of global warming; and, if it even existed. Those who questioned the AGW hypothesis were ignored or discredited. There were no plans for scientific investigations to differentiate anthropogenic from natural causation in prospective global warming such as were initially proposed by Oxford, Cambridge, and the British Royal Society. Politics prevailed over science to pressure scientists to accept the IPCC narrative. From the outset, the IPCC predicted catastrophic man-made global warming and actively promoted the AGW hypothesis through pseudo-scientific activities. Those efforts were designed to prove the hypothesis with a goal towards the elimination of fossil fuel emissions and addressing socio-economic inequalities throughout the world.

The IPCC led the way in efforts to develop sophisticated computer models that were designed to predict changes in global temperatures. Over time, those models became more complex to refine and combine the various calculations involving thermodynamic interactions among the Earth's land mass, atmosphere, and ocean, as well as include several variables dealing with socio-economic factors. At last count, 65 climate models using complex, coupled partial differential equations were in use. These models often involve more than one million lines of computer code and require multiple Cray XC40 super computers to solve the functions. *By the UN IPCC's own admission, these models fail to prove their efficacy through "back testing."* Most importantly, the models have consistently predicted more warming than has actually occurred.

As if the effort to predict future global temperature change was not enough of a challenge, the IPCC incorporated into its models' predictions concerning such variables as projected population growth, change in GDP, land use, energy sources, costs and alternatives, gender inequality, intergenerational inequality and income and asset inequality, all to predict the environmental and socio-economic impact of climate change. Anyone with even an elemental understanding of mathematics would immediately see the futility of such an effort. How could anyone accurately quantify each of these variables and predict their rate of change with any degree of accuracy?

In conclusion, I was surprised to learn that the UN IPCC was permitted to include in its initial purview the analysis of the socio-economic impacts of climate change. Just keeping up with the state of knowledge in climate science with respect to climate change would seem to be enough of a challenge. Now the focus of the UN IPCC seems to be: how can the world eliminate inequality among nations along multiple socio-economic dimensions within the framework that is developed to address climate change? *The answer seems to me to be an effort to develop a mechanism(s) to transfer wealth from wealthier, developed nations, to ones who are not.*

The possible mechanisms could include a limit on mining or producing carbon-based fuels by the developed countries, a tax on carbon emissions and/or a limit on carbon emissions by the developed countries. Such a move would increase the manufacturing costs and/or reduce the manufacturing output of the developed nations and permit developing nations to fill the gap, thereby addressing socio-economic inequalities. Or a carbon tax could be levied on businesses/nations that emit CO_2 above certain pre-determined levels and the proceeds of the tax distributed to the developing nations on some basis (after a share goes to the U.N.). Presumably, the U.N. would collect those taxes and distribute the monies to those countries who had been "disproportionately affected" by climate change, however that may be defined. Another idea that seems to have been discussed in the various U.N. subcommittees that are charged to address the issue is for developed nations to either make payments directly to those nations "disproportionately affected" by climate change, or fund investments in programs in those countries to address climate change and socio-economic inequality, however that may be defined. It seems that all of those "details" would be worked out later, presumably by U.N. representatives.

The Thermodynamic Interactions with the Earth's Land Mass

In Chapter 8, several important facts were established with respect to an analysis of the thermodynamic interactions with the Earth's land mass and the AGW hypothesis. **First, there is no historical database that contains accurate records measuring the surface temperature of the Earth.** The data from the meteorological surface air temperature stations (MSATS) is corrupt because of the sampling technique (the stations are located around 2m above ground and sample surface air temperature) and location, often near urban heat islands which bias temperature readings. In addition, the MSATS data is adjusted for several factors requiring subjective judgment that is subject to bias. The UN IPCC stated that the lack of accurate historical temperature data made back testing of UN IPCC computer climate models that attempt to predict future temperatures difficult, if not impossible. **If a model cannot accurately predict historical results, how can one reason-**

ably assume that it can accurately predict future results?

Second, the concept of an average temperature of the Earth's surface is a meaningless term in science and is an abstraction. Temperature is a thermodynamic property that represents the average kinetic energy of the molecules in a system under analysis. It relates the ability of the molecules in the system to transfer heat energy to molecules in an *adjacent* system with a lower temperature (lower average molecular kinetic energy) via molecular collision. The temperature at any point in time and space in the Earth's landmass is a function of the amount of solar insolation; the albedo of the atmosphere and the Earth's surface; and, local atmospheric conditions including ambient temperature, cloud cover, relative humidity, wind speeds, and ground moisture content. It is different at virtually every point on the globe. Calculating the average temperature of any component of the biosphere is a meaningless exercise.

Third, the concept of an equilibrium temperature of the Earth's surface is a figment of the climate scientist's imagination. The surface temperature is in a constant state of flux as the various thermodynamic interactions occur. Therefore, any effort to determine the Earth's "average energy budget" or average radiative power employing the Stefan-Boltzmann Law is equally invalid. **Absent an accurate equilibrium temperature (which does not exist), any effort to calculate the heat flux from LWIR is merely a theoretical exercise.**

Fourth, the Stefan-Boltzmann Law is used to calculate the radiant heat power of electromagnetic radiation emitted by a body in thermal equilibrium at a given temperature. However, the converse that a change in temperature of the Earth's surface can be calculated by assuming a change in net LWIR flux *absorbed* by the Earth's surface using the Stefan-Boltzmann Law is fallacious reasoning. Climate scientists cannot calculate global warming from a theoretical change in downwelling net heat flux. The Stefan-Boltzmann Law only applies to a system in thermal equilibrium. At the Earth's surface, any change in LWIR flux must be added to the rest of the time dependent flux terms (convection, latent heat, etc.) and coupled to the heat capacity of the land mass. Any resulting change in temperature over a given time must be determined by dividing the change in heat content (enthalpy) by the heat capacity of the local thermal reservoir. **The assumption of an equilibrium average surface temperature was the fundamental climate error made by Arrhenius in 1896 and was incorporated in the climate models when they were first developed in the 1960s. This fact is one of the fundamental flaws of present-day climate science models.**

Fifth, as the result of a flawed understanding of the greenhouse gas effect, many climate scientists today wrongly assume that the primary heat transfer mechanism cooling the Earth's surface is radiant heat transfer, or the emission of LWIR. The fact is that most of the heat transfer is convection heat transfer to the lower troposphere and into the upper layers of the atmosphere. Since the amount of convection heat transfer varies depending upon several factors, and is in constant flux, it is impossible to calculate with any degree of accuracy. None of these values have been or can be accurately measured since they are dynamic.

Sixth, like the concept of Earth's equilibrium temperature, the concept of the Earth's Average Energy Budget is a flawed model based on assumptions concerning estimated parameters that have a very

large range of possible values. Albedo is the proportion of the incident light or radiation that is reflected by a surface. The albedo of the Earth's surface and atmospheric components like clouds significantly influences the amount of energy from the Sun that is absorbed/reflected by the Earth; and that value varies significantly over space and time. As the Russians, Chinese and NASA believe, the volume and location of clouds in the atmosphere plays a very important role in the flow of heat through the atmosphere and to the Earth's surface. Clouds reflect incoming sunlight and absorb LWIR radiated by the Earth. Cloud albedo is estimated to range from 10% to 90%. The cloud volume in the atmosphere is in a constant state of flux and cannot be estimated with any reasonable degree of accuracy.

The following is a statement from NASA concerning the important role that clouds play in Earth's climate: ***"Even small changes in the abundance or location of clouds could change the climate more than the anticipated changes caused by greenhouse gases, human-produced aerosols, or other factors associated with global change*** *(emphasis added)." [15] "Another modeling problem is that* ***clouds change almost instantaneously compared to the rest of the climate system*** *(emphasis added).*" [15]

Therefore, given the extremely important role that clouds play in the Earth's climate and the inability to accurately model that role, how can one possibly develop a computer model that can predict future temperature changes with any degree of accuracy?

In conclusion, as it relates to the thermodynamic interactions with the Earth's land mass, while the efforts to model those thermodynamic interactions might be well intended, the scientific information needed to make those calculations credible does not exist; and, quite frankly, may never exist. The system is just too dynamic, with too many variables that cannot be accurately quantified. At best, given the technology available today, the models require averages, assumptions, and best guesses – hardly the stuff of hard science. **The initial averaging assumptions and radiative forcings used in the climate models mean that the results must be invalid before any algorithms are developed or a single line of code is written.** In 1990, the UN IPCC report predicted that the "global mean temperature" would rise 0.3°C per decade. The temperature databases for the land mass, as flawed as they are, contradict the UNIPCC predictions and depict a *de minimus* temperature rise. **Simply stated, there has been no statistically significant global warming of the Earth's landmass.**

The Thermodynamic Interactions with the Earth's Atmosphere

The AGW hypothesis is based in part on the concept that there must be a short-term energy balance involving the solar irradiance that enters the top of the Earth's atmosphere and that which exits it. This concept is based on the First Law of Thermodynamics, which states that energy can neither be created nor destroyed in an isolated system; known as the conservation of energy principle. Notwithstanding the fact that the Earth is not an isolated thermodynamic system, the assumption implicit in this concept is that if less energy leaves the top of the atmosphere than enters it, global warming must occur. Consequently, the Earth's Energy Budget was developed by climate scientists in an effort to quantify energy flows through the atmosphere and is used as a basis to determine if a theoretical energy imbalance exists.

In Chapter 9, the concept of the Earth's Energy Budget was examined, and it was demonstrated that the assumptions implicit in the development of that theoretical model rendered it meaningless. In addition, the fact that the heat capacity of the Earth's land mass and oceans is ignored in the theoretical construct of the Energy Budget and global warming hypothesis falsifies the analysis.

Another important aspect of the AGW hypothesis deals with the concept that an increase of 120 ppm in the concentration of CO_2 molecules in the lower troposphere "trap" LWIR photons and reradiate those photons back to the Earth's surface, heating the surface and atmosphere, thereby causing global warming. An examination of the behavior of an excited CO_2 molecule employing quantum mechanics demonstrates that the heat energy gained by the absorption of a LWIR photon is transferred by molecular collision to a nearby molecule, not immediately reradiated back to Earth. Further, fundamental atmospheric physics demonstrates that only a small portion of that heat energy involved in the molecular collision is reradiated back down to the Earth's surface; the balance is radiated in other directions or transferred by convection to the upper atmosphere. Based on scientific research employing high resolution spectrographic analysis in conjunction with established radiative heat transfer models, it was demonstrated that **the value of the downwelling heat energy that is reradiated towards the Earth is de minimus, around 2 $W \cdot m^{-2}$, about 0.2% of the maximum solar irradiance that reaches the Earth's surface on a clear summer's day.**

Finally, the UAH satellite-derived temperature dataset for the period 1979 to 2020 has measured a temperature anomaly in the lower troposphere during that period of 0.134°C per decade, or 0.0134°C/yr over a 40-year period. Such an increase is de minimus and arguably within the measurement margin of error. **Therefore, the temperature data-**

base of the troposphere and scientific analysis, using first principles of science, demonstrates there has been no statistically significant warming of the atmosphere and that man-made global warming has not occurred.

The Thermodynamic Interactions with the Earth's Oceans

In 2000, the Argo Float program was established to develop reliable data on certain parameters of the world's oceans at various depths, including temperature, salinity, and currents. The primary purpose of the Argo program was to obtain a base line of data and analyze changes in the ocean's heat content (OTC). *Prior to the Argo program, there was not a reliable historical temperature database for the world's oceans.* Temperature sampling was done on an ad hoc basis, using a variety of sampling, and measuring techniques that changed over the years and produced inaccurate results.

The Argo data available for the period 2000 to 2012 depict an increase in the ocean's average surface temperature of 0.03°C. The average temperature readings at depths from zero to 2000 meters show even less of an increase at 0.02°C. Scientific research using high resolution spectroscopy has demonstrated that downwelling LWIR does not penetrate more than 100 microns below the ocean's surface. **Therefore, an increase in downwelling LWIR emitted by any greenhouse gas, including CO_2, has no effect on the ocean's temperatures.** Any heat gained by the world's oceans can only be the result of solar irradiance. The Argo Float Program temperature database for the world's oceans indicates that there has been no statically significant increase in the temperature or heat content of the oceans. Further, spectrographic, and thermodynamic analyses demonstrate that the oceans cannot be heated by LWIR emissions from the atmosphere.

These analyses prove that there has been no statistically significant warming of the world's oceans. Scientific analysis using proven scientific principles demonstrates that it is not possible for anthropogenic emissions to affect the temperature of the world's oceans.

Sunspot Cycles and Changes in Solar Irradiance Produce Warming and Cooling Periods on Earth

Much of the scientific investigations of historical climate phenomena such as CO_2 concentration in the atmosphere, temperature, and solar activity, is done using proxy analysis; that is, analyzing a variable in a specimen that is believed to be related to a climate phenomenon. One example of such a proxy analysis is the spectrographic analysis of radiocarbon, the carbon 14 isotope (14C) found in plant matter. Scientists believe that an increase in solar activity (sunspots) reduces the amount of 14C in the Earth's atmosphere, which in turn reduces the amount of 14C absorbed by plant matter on the Earth. A reduction in sunspot activity has the opposite effect of increasing 14C in the atmosphere. Scientists employ spectrographic analysis to determine the amount of 14C in tree rings in an effort to quantify and date sunspot activity. [17] Therefore, the quality of the data is dependent on the efficacy of the proxy.

The intensity of the Sun varies along with the 11-year sunspot cycle as depicted in Figure 22. When sunspots are numerous, the solar constant is higher (about 1367 $W \cdot m^{-2}$). When sunspots are scarce, the value is lower (about 1365 $W \cdot m^{-2}$). The solar constant can fluctuate by ~0.1% over days and weeks as sunspots grow and dissipate. [18] Recent research indicates that the change in irradiance varies considerably with the wavelength of the electromagnetic radiation emitted, especially in the ultraviolet, extreme ultraviolet and x-ray spectrums and its resultant effects on the Earth's magnetosphere. [19] Therefore, while there is a direct relationship between changes in sunspot activity and the Earth's climate, there does not appear to be a linear relationship between the magnitude of the changes in the solar constant due to sunspots and the impact on the Earth's climate. Said another way, due to the limited amount of direct, measured historical data regarding sunspot activity and the Earth's temperature, we know that reduced sunspot activity leads to reduced temperatures on Earth, we just don't know the direct correlation between the two. During the period 1645 - 1715 (a period astronomers call the "Maunder Minimum"), the sunspot cycle stopped; the face of the Sun was nearly blank for 70 years. At the same time, Europe was hit by an extraordinary cold spell: the Thames River in London froze, glaciers advanced in the Alps, and northern sea ice increased. An earlier centuries-long surge in solar activity (inferred from isotope studies in tree rings) had the opposite effect: The Medieval Warm Period is thought to have occurred between 950 A.D. and 1250 A.D. While there are no temperature records for the Medieval Warm Period, anecdotal evidence exists that the Earth experienced warmer temperatures. Vikings were able to settle the thawed-out coast of Greenland in the 980s, and even grow enough wheat there to export the surplus to Scandinavia. [12, 13]

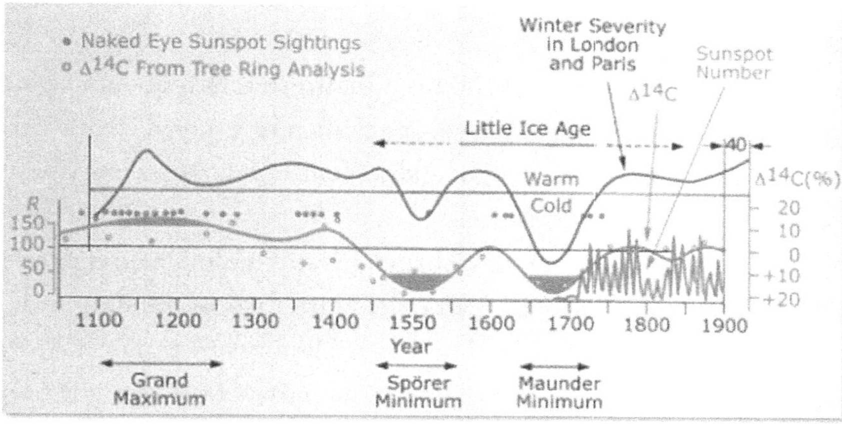

Figure 22. "The Estimated Sunspot Cycle for the period 1000 - 2000 based on observations with the human eye and Tree Ring Analyses." Courtesy of USGS. [19]

In conclusion, the Sun is the heat source for the Earth. The ocean is the thermodynamic body on Earth that determines the Earth's climate, and the Sun drives it. Changes in the amount of solar irradiation that reaches the Earth is the major determinant in the change in the Earth's climate. The sunspot cycle, which is an intrinsic characteristic of the Sun's changing magnetic field over an eleven-year cycle, has a small effect on the value of solar irradiance that reaches the top of the Earth's atmosphere. However, there are three other factors that affect the long-term change in the value of the solar irradiance that reaches the top of the Earth's atmosphere: the elliptical path (eccentricity) of the Earth's orbit around the Sun; the orientation of its axis (obliquity) as it revolves around the Sun; and the rotation about its axis (precession) as it revolves around the Sun. We have the research done by Serbian astrophysicist Milutin Milankovitch in the early part of the twentieth century to thank for this insight. [20]

Changes in the Obliquity and Eccentricity of the Earth's Orbit about the Sun

While the Earth revolves around the Sun, it also rotates about its poles. If the poles were perpendicular to the Sun's rays, everywhere on Earth would have 12 hours of daylight and 12 hours of darkness every day of the year. In fact, it would be hard to mark the passing of a year because there would be no seasons. The changing pattern of night and day and the seasons all come about because the Earth's axis of rotation is tilted about 23.5°off perpendicular (known as obliquity in astronomy) to our orbital plane around the Sun. So, for six months of the year, the north pole is pointing towards the Sun and experiencing summer, and during other six months the Sun lies in the southern hemisphere. Both polar regions of the Earth are cold, primarily because they receive far less solar radiation than the tropics and mid-latitudes due to the spherical shape of the planet. [21]

At either pole the Sun never rises more than 23.5 degrees above the horizon and both locations experience six months of continuous darkness. Moreover, most of the sunlight that does shine on the polar regions is reflected by the bright white surface of ice and snow. Two annual (December and June) solstices are the times when the Earth is at the point in its orbit that the poles are most tilted towards the Sun. Because one pole is bathed in 24-hour daylight at this time, it receives more sunlight than anywhere on the equator ever does – but most is reflected straight back into the atmosphere by ice and snow. [21]

The Serbian astrophysicist Milutin Milankovitch is best known for developing one of the most significant theories relating Earth motions and long-term climate change. Born in 1879 in the rural village of Dalj (then part of the Austro-Hungarian Empire, today located in Croatia), Milankovitch attended the Vienna Institute of Technology and graduated in 1904 with a doctorate in technical sciences. After a brief stint as the chief engineer for a construction company, he accepted a faculty position in applied mathematics at the University of Belgrade in 1909 — a position he held for the remainder of his life. [22]

Milankovitch dedicated his career to developing a mathematical theory of climate based on the seasonal and latitudinal variations of solar radiation received by the Earth. Now known as the Milankovitch Theory, it states that as the Earth travels through space around the Sun, cyclical variations in three elements of Earth-sun geometry combine to produce variations in the amount of solar energy that reaches Earth: [22, 23]

1. Variations in the Earth's orbital eccentricity—the shape of the orbit around the Sun.
2. Changes in obliquity—changes in the angle that Earth's axis makes with the plane of Earth's orbit. (Figure 24)
3. Precession—the change in the direction of the Earth's axis of rotation, i.e., the axis of rotation behaves like the spin axis of a top that is winding down, hence it traces a circle on the celestial sphere over a period of time. (Figure 24)

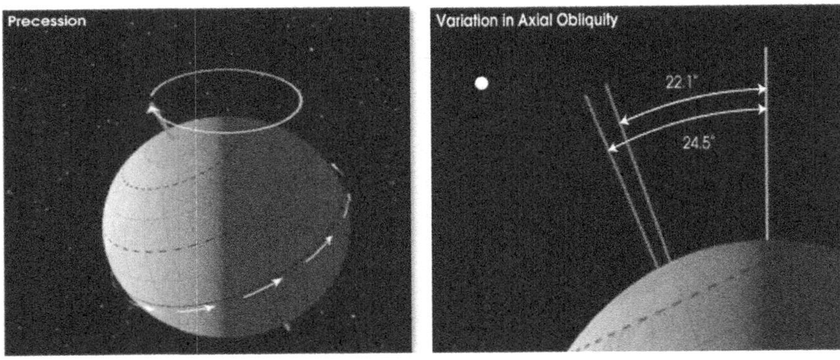

Figure 23. Milankovitch's Theory. Courtesy of NASA. [23]

For about 50 years, Milankovitch's theory was largely ignored. Then, in 1976, a study published in the journal "Science" examined deep-sea sediment cores and found that Milankovitch's theory did correspond to periods of climate change. [24] Specifically, the authors were able to extract the record of temperature change going back 450,000 years and found that major variations in climate were closely associated with changes in the geometry (eccentricity, obliquity, and precession) of Earth's orbit. Indeed, ice ages had occurred when the Earth was going through different stages of orbital variation. In conclusion, the amount of energy that the Earth receives from the Sun - the solar constant - is not actually constant, but varies over time because of the changes in geometry of the Earth's orbit about the Sun. The mean solar constant has been measured by satellites to be about 1365 $W \cdot m^{-2}$. However, that value can vary by as much as +/- 3% (41 $W \cdot m^{-2}$). [18] This amount far exceeds any changes in the solar energy that reaches the Earth as a result of sunspots, purported man-made activities or any other known factor.

Since this study, the National Research Council of the U.S. National Academy of Sciences has embraced the Milankovitch Cycle model: *"orbital variations remain the most thoroughly examined mechanism of climatic change on time scales of tens of thousands of years and are by far the clearest case of a direct effect of changing insolation on the lower atmosphere of Earth"* (National Research Council, 1982). [23]

CONCLUSION

SO, WHAT CAN ONE conclude about the AGW hypothesis based on a careful investigation of the facts concerning each of the underlying claims? **It is a fraudulent hypothesis developed to promote climate change research funding and to advance a socio-economic agenda. It has no basis in scientific principles or facts.** There has been a *de minimis* increase detected in the temperature of the Earth's troposphere during the last 40 years during which it has been measured by satellites. The world's oceans have not "stored" 90% of anthropogenic warming (there has been none). Likewise, the temperature of the Earth's landmass has not changed by a significant amount. Little, if any, research has been done to distinguish between natural causation and prospective man-made changes with respect to the small temperature increase. Scientific research has confirmed that an increase in the CO_2 concentration in the troposphere has no measurable effect on the Earth's climate. IPCC models are unable to accurately predict the past or future temperature changes in the Earth's biosphere. **The true goal of the UN IPCC is to "eliminate inequality among nations along multiple socio-economic dimensions" through a massive wealth transfer program from the developed nations to the undeveloped nations.**

Is it possible that the Earth's climate is growing warmer or colder?

Yes. Is man responsible for the change? No. If not man, then who or what? Critical thinking, based on scientific evidence and deductive reasoning, suggests that changes in the eccentricity, obliquity, and precession of the Earth's orbit, in conjunction with changes in solar irradiance due to sunspots, affects the thermodynamic interactions of the world's oceans and creates the Earth's climate. It would seem that future climate research should focus on these issues in an effort to develop a reasoned, unbiased view of what factors affect the Earth's climate.

A Way Forward

Perhaps it is time to convene a world conference similar to the past Solvay conferences on the subject of the science of global climate. The world's preeminent experts in atmospheric physics, physical oceanography, spectroscopy, thermodynamics, atmospheric chemistry and related fields would be invited to attend to discuss the issues pertinent to the development of the Earth's climate. Subjects would include the physics and thermodynamics governing the absorption of a LWIR photon by molecules in the Earth's atmosphere and the absorption of sunlight by the world's oceans; the effects of thermodynamic cycles in the world's oceans such as the Atlantic Multi-decal Oscillations and the Pacific decadal oscillations on the temperature of the lower troposphere and the formation of ice at the poles. Additional subjects would include changes in solar irradiance caused by changes in the eccentricity, obliquity and precession of the Earth's orbit; the sunspot cycle and its effect on the Earth's climate; and the dynamic surface energy transfer processes that affect the temperature of the lower troposphere.

One would expect that the results of such a meeting would be the development of a set of rigorous scientific studies led by these experts, employing the scientific method, to uncover facts pertinent to each scientific topic and its relation to the Earth's climate. The facts developed could then be used to build hypotheses about what occurrences in nature affect the Earth's climate. Then, those hypotheses could be tested through further research and experimentation to either falsify or validate the hypothesis.

Concurrent with scientific studies on factors that may affect the Earth's climate, studies should be conducted on the economic and social effects on the world's nations that might result from the elimination of the use of fossil fuels as an energy source. It is time that the carbon atom receive a fair hearing. In addition, engineering studies should be conducted on the impact of the integration of alternative energy sources such as wind and solar power into electric power grids throughout the world and its prospective impact on the reliability of the grid under certain conditions. *In conclusion, there are no beneficiaries of a rush to judgment regarding important matters of science. Scientific investigation must conform to the scientific method of inquiry to develop objective conclusions and minimize errors in the conclusions that are reached from such investigations.*

The Triumph of Dollars and Politics Over Science: Why You Should Care

In this book, I have criticized the pseudo-science that is employed to support the claims of the proponents of man-made global warming and compared it to established, first principles of science, to falsify the hypothesis. I believe it is important to reveal the truth about the subject: the bias and lack of scientific rigor that characterizes the pseudo-scientific investigations and the motivation for promoting the fradualent hypothesis. However, there is an equally important topic that is associated with this subject: the use of pseudo-science in an effort to manipulate public policy. The question is: *"Why should you care?"*

For most of history, the conduct of science has been the purview of scientists who pursued research based on the scientific method of inquiry for the purposes of advancing mankind's knowledge of the physical world. While kings, dictators and presidents have made occasional forays into directing scientific research with regards to developing weapons of war, or more recently, a vaccine for the coronavirus, the direction and outcome of scientific research has mainly been left to scientists. However, in the last several years that seems to have changed.

The question is why *should* any of us care about the employment of pseudo-science to a subject such as global warming? Failed hypotheses in science have come and gone over the centuries, with no apparent harm to society. Sooner or later, the evidence will demonstrate that the Earth is not on a path to thermal destruction; that is, the Earth will not perish in 2031 due to global warming, as stated by Congresswoman Alexandria Ocasio Cortez in January 2019. In February 2021, Climate Envoy John Kerry made a statement that shortened Cortez's prediction of the Earth's demise by one year, to 2030.

Why do the U.N. and certain politicians in the U.S. and Western Europe continue to promote the fraudulent global warming hypothesis? With regards to the U.N., it is all about the money – money for research and to affect socio-economic change by transferring wealth from developed nations to developing nations. For many politicians, it is all about the power – the power to control the lives of the electorate. Power is the ultimate narcotic. And money follows power in politics. Suppose that I told you in December 2019, that within months, local and State governments in the U.S. would close schools, dictate which businesses could operate and limit their hours of operation, dictate when houses of worship could open and limit the number of those allowed to attend, and issue regulations that governed social conduct and severely restricted freedom of assembly, all in response to a crisis – the onset of the coronavirus pandemic? You would have rightly stated it could not happen; that all the aforementioned would be an assault on the individual liberties of all Americans that are protected under the U.S. Constitution. You would probably say that if any government should attempt such actions, the courts would step in to invalidate them. However, just those very series of events occurred.

Now, suppose I told you that in the near future, the government would mandate what had previously been free-market decisions about the sources of power generation (coal, gas, oil, solar, hydro, nuclear, wind) in the U.S., as well as the composition of the U.S. power grid, all in an effort to avert a supposed crisis: climate change (formerly known as "global warming")? In addition, that the federal government would usurp the role of the consumer in choosing what energy sources would power homes, businesses and factories; how automobiles should be powered (electric vs. internal combustion engines); and ultimately, what modes of transportation (automobile vs. public transport) the public could use. President Biden and Climate Envoy Kerry have declared climate change a crisis and an "existential threat" to the world. Seem farfetched? I don't think so.

The fact is that the scientific rationale that advised the response to the coronavirus pandemic has changed over the last year as more facts emerged about the virus: how it is transmitted and how it should be treated and contained. Contradictory evidence has emerged regarding the effectiveness of social "lockdowns" in limiting the spread of the virus, the transmission of the virus from surfaces and human contact, and even the origin of the virus itself. I believe this demonstrates the danger of government at any level rushing to judgment and misusing "science" to justify the erosion of civil liberties. The erosion of civil liberties is a slippery slope. Once it begins in response to a crisis or an "existential threat," it can be difficult to stop. Who defines what is truly a crisis or an existential threat? Who decides how to respond and what that response is based upon? Is it based on the "science *de jour?*"

Rahm Emmanuel, former President Obama's White House Chief of Staff, once famously stated "You never want a serious crisis to go to waste. And what I mean by that is an opportunity to do things that you think you could not do before."

On January 27, 2021, President Joe Biden signed an executive order which stated the following [26]: *"President Biden set ambitious goals that will ensure America and the world can meet the urgent demands of the* ***climate crisis*** (emphasis added), *while empowering American workers and businesses to lead a clean energy revolution that achieves a carbon pollution-free power sector by 2035 and puts the United States on an irreversible path to a net-zero economy (sic) by 2050."*

If one takes this statement at face value, it seems that the intent of President Biden is to mandate the elimination of carbon-based fuels power production by 2035. According to the U.S. Energy Information Administration, in 2020, about 60% of the electricity generated in the U.S. came from plants that burned fossil fuels; 20% from nuclear generation; and 20% from "renewable energy sources," such as wind turbines and solar cells. The above statement also suggests that by

2050, the U.S. economy would achieve a "net zero" status, presumably meaning that either CO_2 emissions would be eliminated, or only equal to that which was absorbed by carbon sinks; however, that might be measured.

It would be one thing if the preponderance of scientific evidence demonstrated that CO_2 emissions had a harmful effect on the world's environment (i.e., the "science *was* settled"); but that clearly is not the case. There are many scientists around the globe who dispute that view and the temperature database supports their position. In my opinion, an effort to substantially reduce or eliminate the use of fossil fuels and replace fossil fuel power generation in the U.S. with wind and solar production, based on today's technologies, could have a devastating effect on the economy, the reliability of the power grid, the cost of energy and the standard of living of every American.

The push by the federal government to contain the coronavirus pandemic and the push to eliminate fossil fuels have one thing in common: they both attempt to misuse science to infringe upon individual freedoms and institute fundamental changes in the economy and behavior of individuals in the U.S. The idea that the government could dictate when a house of religious worship could operate and how many people could attend those services would be anathema to most freedom-loving Americans, *except in time of crisis;* that is, during the coronavirus pandemic. Therefore, if one posits climate change as a crisis-an existential threat- it might arguably justify the idea that the federal government could dictate to Americans how electricity is generated in the U.S. or what type of fuel could be used in their businesses, homes, or automobiles, all in the cause to save the planet. In such a situation, science would be sacrificed on the altar of political opportunism.

So, why should every American care about the scenario where money and politics win out over real science? *If pseudo-science becomes the weapon of politicians and regulators; and, it is used to enact extreme policies, then civil liberties can and will be sacrificed under the false façade of the common good.*

Therefore, I believe that it is critical that every American citizen who is concerned about the politicization of the pseudo-science associated with the man-made global warming issue contact their local, state, and federal government representatives about this matter. All concerned should express their opposition to the actions proposed by the proponents of the AGW hypothesis which include attempts to phase out fossil fuels, tax carbon emissions and restructure the US power gird employing a greater percentage of alternative energy sources. Our political leaders should be made aware that the science is not settled on this important matter and that scientific research on the subject needs to be conducted in an unbiased manner based on the scientific method. Any efforts to abridge the personal freedoms guaranteed to all Americans under the US Constitution under the guise of mitigating man-made global warming need to be vigorously opposed in the halls of government and in the US courts.

Appendix

History Confirms
Democrat's 1988 Senate Global Warming Hearing Got Everything Wrong from Start to Finish

Guest essay by Larry Hamlin

The June 23, 1988 Democratic Senate Committee on Energy and Natural Resources hearing opened the door on climate alarmism in the nation with testimony from scientific "experts" and Committee Senators who offered speculation and conjecture on a host of weather and climate topics while sharing their scientifically unsupported and sensationalized doomsday perspectives. The complete record of the hearing's proceedings can be found here.

This hearing is often celebrated by climate alarmists and cited as a milestone in establishing the alleged legitimacy of greenhouse gas emissions as the principal cause of increased global warming that also drives other global climate conditions. The hearing was front page headline news in the New York Times.

"All the News That's Fit to Print"

The New York Times

Late Edition

New York: Today, sunny, cool. High 74-79. Tonight, increasing clouds. Low 57-65. Tomorrow, morning clouds, then windy and warmer. High 79-88. Yesterday: High 87, low 67. Details, page A18.

VOL. CXXXVII . . . No. 47,546 | NEW YORK, FRIDAY, JUNE 24, 1988 | 30 CENTS

Global Warming Has Begun, Expert Tells Senate

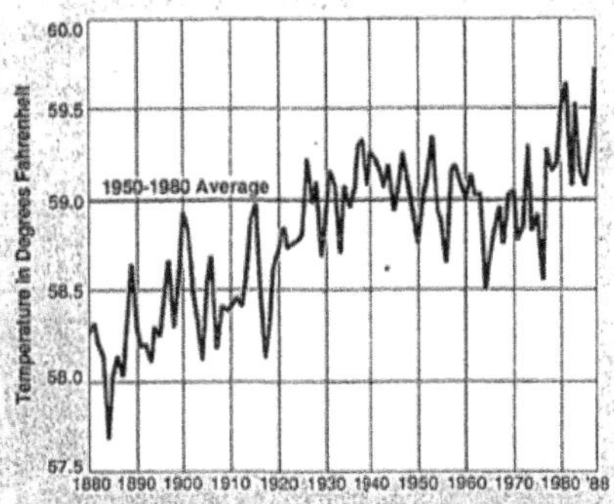

Global Warming: Greenhouse Effect?

Average global temperatures through the first five months of 1988. As a baseline, scientists use the global average from 1950 to 1980.

Source: James E. Hansen and Sergej Lebedeff

The New York Times / June 24, 1988

Sharp Cut in Burning of Fossil Fuels Is Urged to Battle Shift in Climate

By PHILIP SHABECOFF

Special to The New York Times

WASHINGTON, June 23 — The earth has been warmer in the first five months of this year than in any comparable period since measurements began 130 years ago, and the higher temperatures can now be attributed to a long-expected global warming trend linked to pollution, a space agency scientist reported today.

Until now, scientists have been cautious about attributing rising global temperatures of recent years to the predicted global warming caused by pollutants in the atmosphere, known as the "greenhouse effect." But today Dr. James E. Hansen of the National Aeronautics and Space Administration told a Congressional committee that it was 99 percent certain that the warming trend was not a natural variation but was caused by a buildup of carbon dioxide and other artificial gases in the atmosphere.

An Impact Lasting Centuries

Dr. Hansen, a leading expert on climate change, said in an interview that there was no "magic number" that showed when the greenhouse effect was actually starting to cause changes in climate and weather. But he added, "It is time to stop waffling so much and say that the evidence is pretty strong that the greenhouse effect is here."

If Dr. Hansen and other scientists are correct, then humans, by burning of fossil fuels and other activities, have altered the global climate in a manner that will affect life on earth for centuries to come.

Dr. Hansen, director of NASA's Institute for Space Studies in Manhattan, testifed before the Senate Energy and Natural Resources Committee.

Some Dispute Link

He and other scientists testifying before the Senate panel today said that projections of the climate change that is now apparently occurring mean that the Southeastern and Midwestern sections of the United States will be subject to frequent episodes of very high temperatures and drought in the next decade and beyond. But they cautioned that it was not possible to attribute a specific heat wave to the greenhouse effect, given the still limited state of

Continued on Page A14, Column 3

In reality the hearing's climate alarmist statements and claims represented nothing but conjecture and speculation driven by the political ambitions of politicians and scientists seeking fame and additional government funding. The hearing failed to address scientifically proven and verifiable climate evidence.

For more than three decades since this hearing the Democratic Party has continued to engage in scientifically unsupported climate alarmism and relied instead upon conjecture, speculation and exaggeration while concealing the failure of climate model projections and more importantly the UN IPCC acknowledgment that climate models cannot provide accurate assessments of future climate states.

A detailed review of the statements and claims made by both the Senators and scientific "experts" presenting greenhouse gas and global warming information at the hearing when viewed in 2021 after 33 years of recorded climate data reveals how extraordinary flawed and mistaken the hearings proceedings were with the numerous failed claims being completely ignored by the main-stream media that continues to celebrate the climate alarmism symbolism of this hugely inaccurate and misleading hearing.

The incredibly flawed perspectives of this hearing rather than being celebrated as a major milestone of climate alarmism success instead represent everything that is wrong about climate alarmism's use of politically motivated and contrived speculation, conjecture, exaggeration, distortion, and deception in making scientifically unsupported claims in addressing climate issues.

The hearing was held in the Senate Dirksen Office Building in Washington D.C. on a hot 101 Degree F day during an unseasonably warm heat wave (the highest temperature on record in Washington D.C. occurred on July 29, 1930 at 106 degrees F some 91 years ago) along with a major drought also underway in the Midwest and Southeastern regions of the country.

The incredibly flawed perspectives of this hearing rather than being celebrated as a major milestone of climate alarmism success instead represent everything that is wrong about climate alarmism's use of politically motivated and contrived speculation, conjecture, exaggeration, distortion, and deception in making scientifically unsupported claims in addressing climate issues.

HEARING

BEFORE THE

COMMITTEE ON ENERGY AND NATURAL RESOURCES UNITED STATES SENATE

ONE HUNDREDTH CONGRESS

FIRST SESSION

ON THE

GREENHOUSE EFFECT AND GLOBAL CLIMATE CHANGE

JUNE 23, 1988

PART 2

The hearing was Chaired by Democratic Senator Bennett Johnston of Louisiana whose opening statement introduced the primary hearing topics to be addressed which were the greenhouse effect and global warming and their impact on global climate.

Those presenting at the hearing included a number of Senators along with NASA scientist Dr. James Hansen, Environmental Defense Fund Senior Scientist Dr. Michael Oppenheimer, Dr. George Woodwell, Director of Woods Hole Research Center and others as noted below.

CONTENTS

STATEMENTS

APPENDIXES

APPENDIX I

APPENDIX II

(III)

Some regions of the country were experiencing drought in 1988 that was significantly reducing production levels of soybeans, cotton, corn crops and other agricultural products. This drought became a major point of discussion at these hearings with presenter after presenter speculating devoid of any rational support that this droughts occurrence and severity were probably caused by increased greenhouse gases without offering any confirming scientific evidence regarding such claims.

The information provided below follows the course of the hearing from opening statement through subsequent presenters in order of appearance highlighting key portions of the presentations with comparisons using updated climate data since the hearing to assess the scientific legitimacy or lack thereof regarding the hearing's climate claims. As arduous as this task is it is necessary to provide a clearer understanding of the hearing's scientific ineptness and politically driven climate alarmist bias.

Conclusions are provided as noted in bold face type at the end of key discussion topics.

Senator Bennett opened the hearing by noting:
"The greenhouse effect has ripened beyond theory now. We know it is a fact. What we don't know is how quickly it will come upon us as an emergency fact, how quickly it will ripen from just a matter of deep concern to a matter of severe emergency.

And what we don't know about it is how we're going to deal with it and how we're going to get the American people to understand that perhaps this drought which we have today is not just an accidental drought, not just the kind of periodic drought which we have from time to time but is, in fact, the result of what man is doing to this planet."

In his prepared statement Senate Bennett further noted:

"The current drought situation teaches us how important climate is to the nation's social, economic, and physical well-being. The United States is currently mobilizing its political and financial resources to grapple with the enormous agricultural devastation of the present dry spell over the Midwest and Southeast portions of the United States. The present drought graphically illustrates only a small portion of the scenario which could transpire if global warming and climate change predictions are accurate."

"Taking the proper steps to control the degree and pace of global warming will not be easy."

"Nevertheless, the United Sates must take a concerted effort to increase its use of energy sources that emit relatively less carbon dioxide and other trace gases."

Based upon this statement Senator Bennett had clearly tipped the Democrats hand by announcing in his opening remarks a conclusion that recent global warming is caused by increasing greenhouse gas emissions and as a consequence actions must be taken immediately to reduces these gases even before any confirming studies and analysis have even been identified much less undertaken.

Senator Bennett then turned the meeting over to Democratic Senator Wirth of Colorado to conduct the meeting. Further amplifying the greenhouse gas caused 1988 drought climate speculation Senator Wirth noted:
"In the past week, many of us have been seeing first-hand the effects of the drought that is occurring across the heart of this country. Meteorologists already are recording this as the worst drought this nation has experienced since the Dust Bowl days of the 1930s. The most productive soils and some of the mightiest rivers on earth are literally drying up."

"Already, more than 50 per cent of the Northern Plain's wheat, barley, and oats have been destroyed and the situation could get much worse."

"We must begin to ask: is this a harbinger of things to come? Is this the first greenhouse stamp to leave its impression on our fragile global environment? I understand that Dr. Hansen will provide testimony that points clearly in that direction."

A question asked by Republican Senator McClure of Idaho during the Q&A session of the hearing requested more scientific substance regarding the claimed connection between greenhouse gases and droughts with the response clearly exposing the speculation, conjecture and limitations of the climate "science" being addressed at the hearing.

Senator McClure:
"But that doesn't explain the droughts of the 1930s. Was the drought in the middle 1930s a result of the greenhouse effect?"

Dr. Hansen:
"You will notice in the climate simulations which I presented we began the simulations in 1958. That was the international geophysical year. The measurements of atmospheric composition began at that time and have been accurate since that time."

"It is more difficult to go back and simulate the 1930s because we don't know what caused the 1930s to be warmer than the preceding decades. So, it is really difficult to say what caused the droughts in the 1930s."

Regarding the numerous drought and heat wave claims made by Senators Bennett and Wirth as noted above analysis of climate information that has occurred since the hearing provides a much clearer picture of real-world heat wave and drought outcomes.

With the passage of 33 years since the Democratic Party's politically contrived greenhouse effect and global warming hearing it is now appropriate to look at what actually happened regarding the occurrence of droughts both in the U.S. and globally using more than three decades of recorded data versus the climate alarmism speculation and conjecture expressed at the 1988 hearing concerning the mid 1980s drought as well as the occurrence and severity of future droughts in the U.S.

EPA data shows no increasing trend in the occurrence and severity of droughts in the U.S. over a period of more than a century with the droughts of the 19030s and 1950s clearly remaining the most extreme on record and dominating drought severity compared to the droughts of the 1980s and droughts of other decades as well.

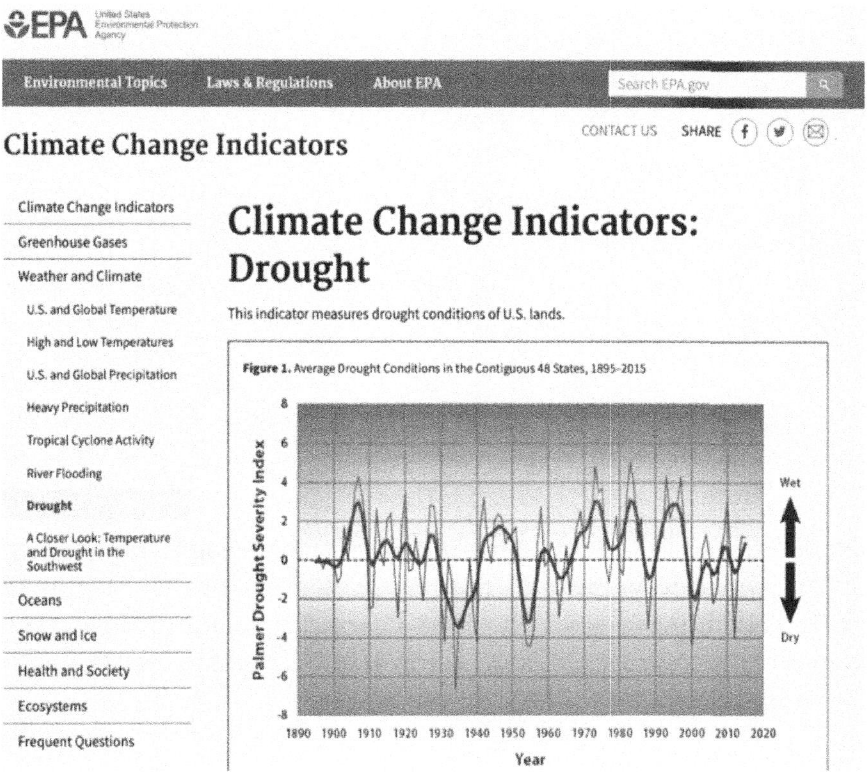

EPA data also shows no trend of increased heat waves across the U.S. with the 1930s overwhelmingly dominating the occurrence of heat waves for a period of more than a century.

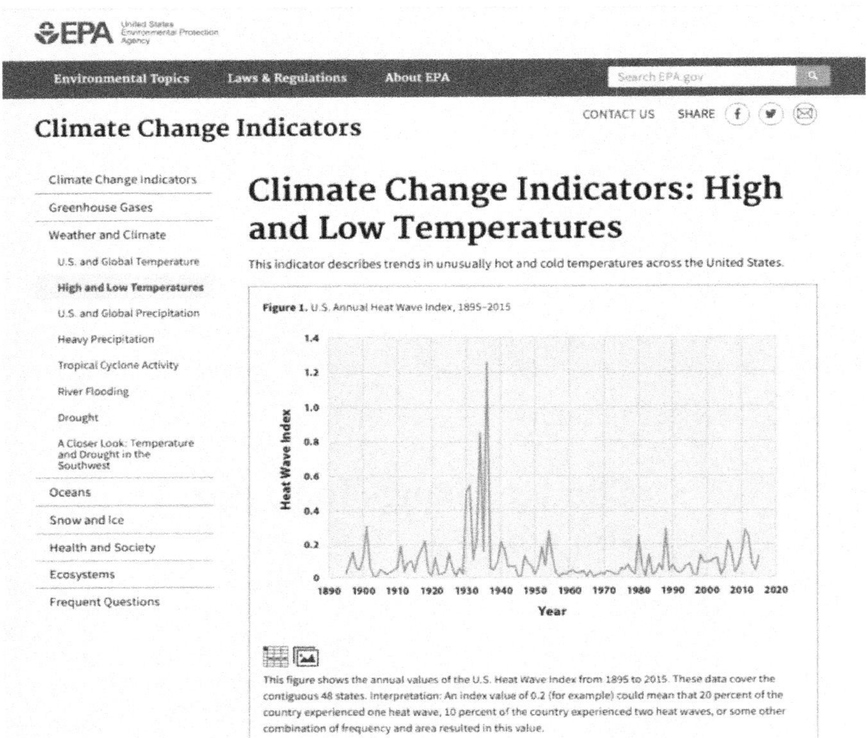

NOAA data on droughts shows no global increasing trend of severity of global droughts with the IPCC's decades of regular climate reviews conducted since the 1988 hearing concluding "low confidence" in increasing global heat waves and drought severity due to the impact of increased greenhouse gases.

GWPF Factsheet 1

GWPF

Drought

Are droughts getting worse on a global scale?

The IPCC says it is hard to say ('low confidence') whether global drought has become better or worse since 1950.[1]

Figures from the US National Oceanic and Atmospheric Administration (NOAA) show no trend in the proportion of the globe in drought since 1950 (see Figure 1).[2] Others have suggested a decline in drought levels in recent decades.[3]

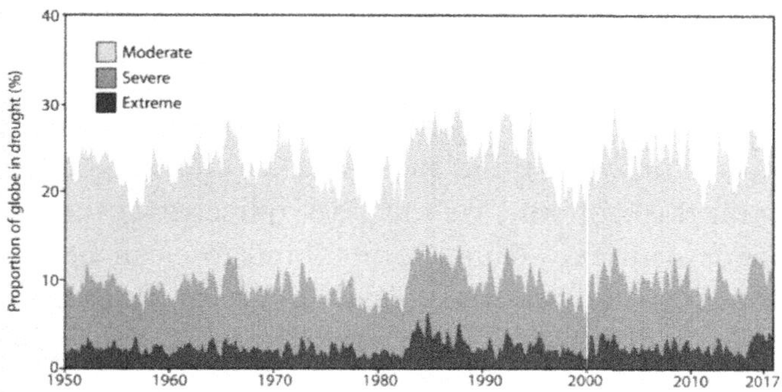

Figure 1: The proportion of the globe in drought is largely unchanged.

What about regionally?

The lack of a global trend disguises changes that have been happening regionally. Some regions have become dryer, and others wetter. The IPCC has noted drought increases in the Mediterranean and West Africa but drought decreases in central North America and northwest Australia.[3]

History has shown that the conjecture and speculated expressed at the Senate hearing between greenhouse gas emissions and the increased occurrence and severity of U.S. droughts and heat waves as well as increasing global droughts is unsupported by more than three decades of recorded global climate data that occurred after the 1988 hearing.

The Democratic Party's hearing with their exaggerated climate alarmist claims speculating greenhouse gas emissions impacts being linked to the 1988 drought as well as likely causing future increased numbers and severity of droughts and heat waves have been proven WRONG.

The next presenter at the hearing offering views on greenhouse gases and their connection to global warming was Democratic Senator Max Baucus of Montana a long-time member of the Committee on Environment and Public Works.

The Statement of Senator Baucus noted the following key points:
"Like those who believe the stock market crash of October was a warning on the economy, we must ask ourselves if the drought we are facing is nature's warning to mankind to clean up its act." (As noted previously the hearings speculation regarding the occurrence and severity of droughts have proven to be WRONG.)

"The inhabitants of planet earth are quietly conducting a gigantic environmental experiment."

"The experiment in question is the so-called greenhouse effect – the gradual warming of our atmosphere caused by an overload of carbon dioxide and other trace gases."

"The projected increases in the greenhouse gases are predicted to cause unprecedented global and regional climate changes."

"Temperature will increase. Current models predict an increase in average global temperature of 1.5 to 4.5 degrees C by year 2030. That is an increase of about 3 to 9 degrees F in only 40 years." (The same future global average temperature anomaly increase claims are repeated numerous times by presenters during the course of the hearing.)

“Sea level could rise from one to four feet, inundating our coastlines and contaminating drinking water supplies with salt water.” (Presumably over the next 40 years as addressed in his temperature increase comments.)

“We are talking about a situation where mankind has finally wrestled control of the planet from Nature.” (This sounds like a quote from a 1950s science fiction movie.)

“If emissions continue on their present track, we will have committed Earth to a warming of 1.5 to 4.5 degrees C by 2030.”

The global temperature increases noted by Senator Baucus (along with other presenters) were derived from the NASA GISS climate model presented by Dr. James Hansen.

Figure 3 from the testimony of Dr. Hansen (Attachment A Figure 3) shown below addresses the climate model analysis of the 5-year running mean global average temperature anomaly for the period 1986 to 2060 for three greenhouse gas scenarios that were generated by the NASA GISS climate model discussed in more detail in Attachment A to Dr. Hansen's testimony.

Scenario A (solid top line extending to 2060) represents continued increased growth of greenhouse gas emissions and shows that from year 1986 an increased temperature anomaly of about 4.5 degrees C during that period.

Scenario B (top dotted line extending to about 2030) represents continuing greenhouse gas emissions at levels consistent with 1986 emissions levels and shows an increase temperature anomaly of about 1.5 degrees C during that period.

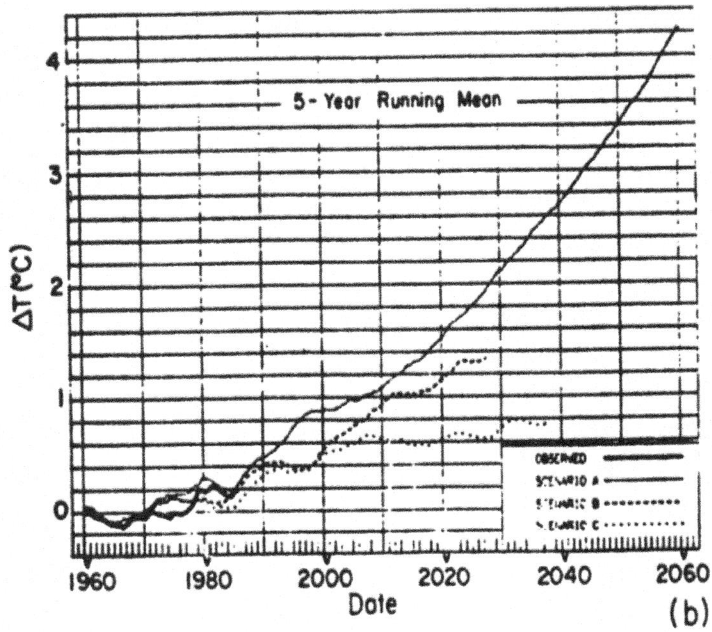

Fig. 3. Annual mean global surface air temperature computed for scenarios A, B and C. Observational data is from *Hansen and Lebedeff* [1987, 1988]. The shaded range in part (*a*) is an estimate of global temperature during the peak of the current and previous interglacial periods, about 6,000 and 120,000 years before present, respectively. The zero point for observations is the 1951–1980 mean [*Hansen and Lebedeff*, 1987]; the zero point for the model is the control run mean.

Scenario C (bottom dotted line extending to about year 2040) represents significantly reduced emissions and shows an increased temperature anomaly of about 0.8 degrees C during that period starting in 1986.

The latest UAH satellite measured global average temperature anomaly for the period from 1979 to 2021 shows a decadal rate of increase of about 0.14 degrees C. For the period staring from 1986 to 2021 the rate of UAH global average temperature anomaly is also about 0.14 degrees C per decade which results in an increase of about 0.49 degrees C during this period.

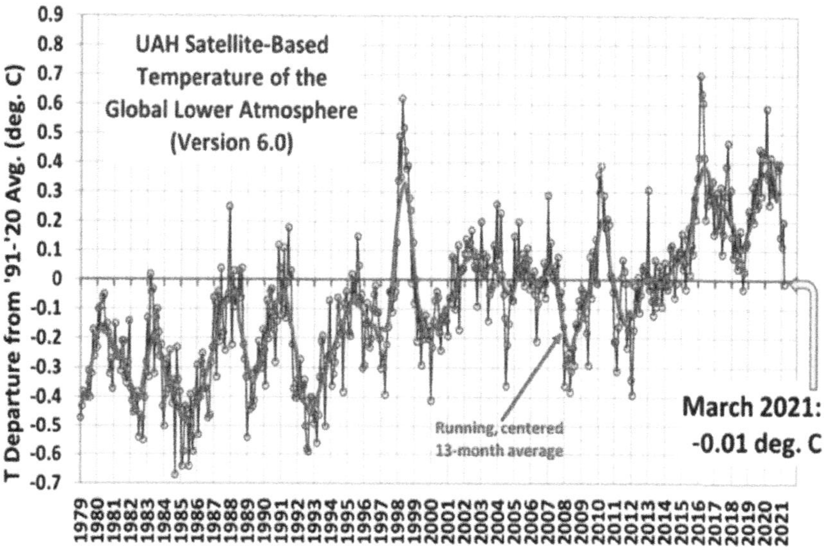

The UAH 0.49 degrees C measured temperature anomaly increase from 1986 to year 2021 is far below the temperature anomaly increases represented by the NASA GISS model used in the Democrats 1988 hearing which shows temperature increases from 1986 to 2021 as being about 1.4 degrees C increase for Scenario A, about 1 degree C increase for scenario B and about 0.6 degrees C for Scenario C.

The Scenario A greenhouse gas emission case is closest to what has happened to global emissions since 1988 with continuing rapid growth **in emissions driven by the world's developing nations led by China and India that now account for 65% of all global emissions as well as all global emissions increases since about year 2006.**

Global emissions increased 67% from 1988 to 2019 with the developing nations completely dominating this increase. U.S. emissions have declined since 2007 with reductions totaling about 1 billion metrics tons through 2019.

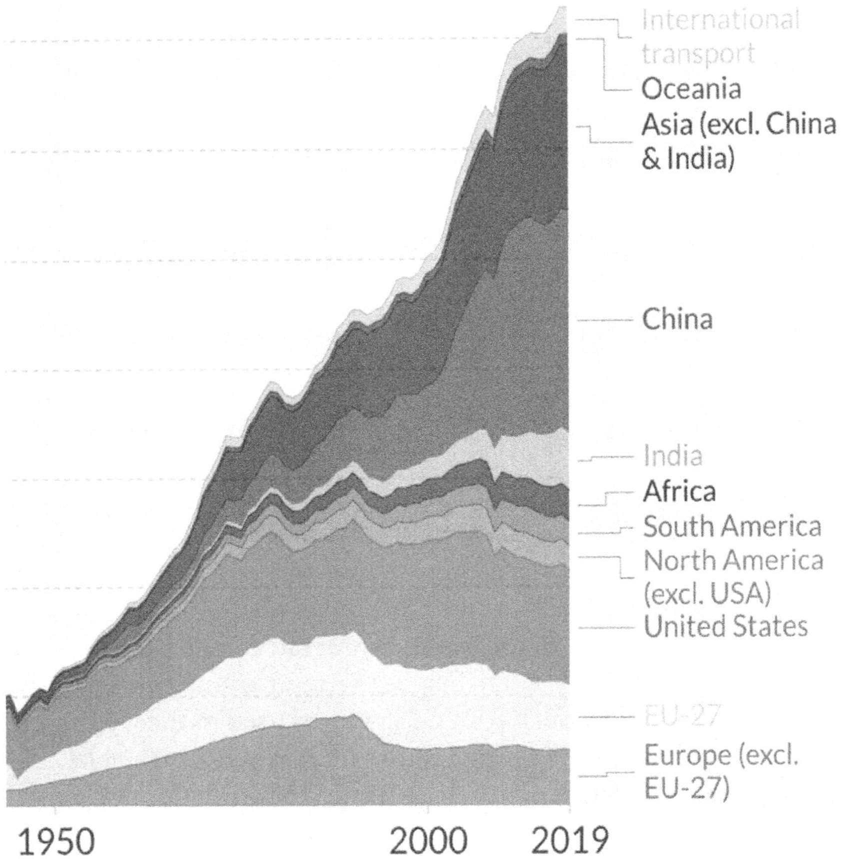

OurWorldInData.org/co2-and-other-greenhouse-gas-emissions • CC BY

The UAH temperature anomaly increase of about 0.49 degrees C in the period between 1986 and 2021 is even below the Scenario C NASA GISS case (0.6 degrees C) which assumed significant reductions in global emissions that never happened.

Additionally the UAH annual global average temperature anomaly has plateaued since the year 2016 El Nino driven peak with trivially insignificant changes (hundredths of a degree C) in the annual global average temperature anomaly during this nearly 6 year-long global temperature hiatus period despite continuing large increases in global greenhouse gas emissions during this 6 year period.

HIATUS IN GLOBAL WARMING

The New Pause lengthens by three months to 5 years 10 months

2 days ago Guest Blogger

202 Comments

By Christopher Monckton of Brenchley

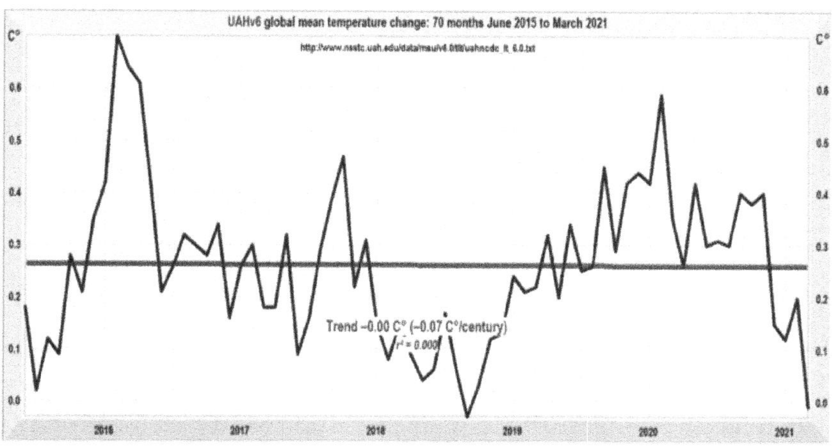

Now that the small La Niña that has recently ended has begun to have its effect on global temperatures, the UAH monthly global mean lower-troposphere anomalies now show a further sharp drop, lengthening the New Pause by three months, from 5 years 7 months last month to 5 years 10 months this month.

Measurements of the global average temperature anomaly since the Democrats 1988 hearing clearly shows the NASA GISS climate model relied upon in these hearings grossly overstates the impact of greenhouse gas emissions on global temperatures.

The claims made by numerous Democratic Party Senators and scientific "experts" about global average temperature anomalies increasingly dramatically in the future because of increasing greenhouse gas emissions highlighted at this hearing (1.5 to 4.5 degrees C temperature anomaly increase by 2030) have been proven WRONG.

Senator Baucus claimed that because of greenhouse gas driven increasing global temperatures sea level could increase and rise by an additional one to four feet over the next 40 years.

NOAA tide gauge data has been updated to include measurements through year 2020 for hundreds of coastal tide gauges stations located around the U.S., Alaska, Hawaii and various other Pacific and Atlantic Island groups.

The longest recorded period of NOAA tide gauge measurements is the station located at the Battery in New York on the U.S. east coast which has been measuring coastal sea level rise at that location since 1856 some 164 years ago.

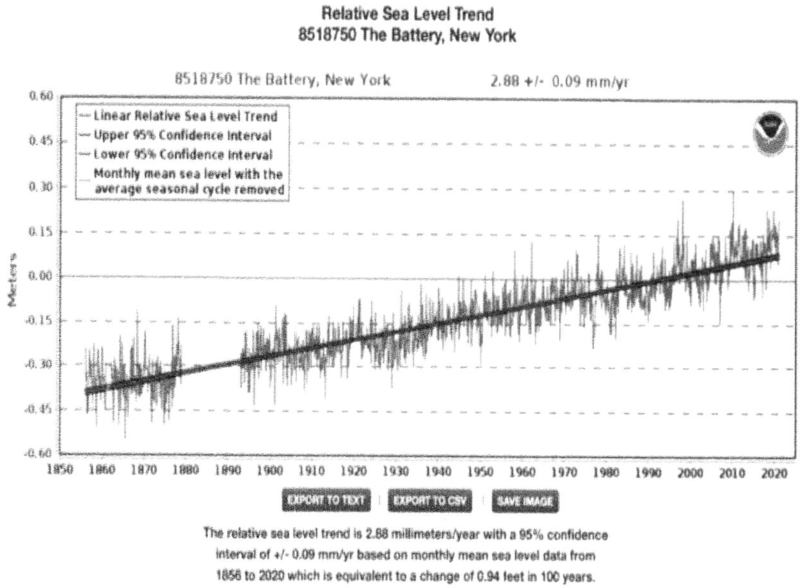

The rate of coastal sea level rise is 0.94 feet per century or 2.88 millimeters per year and has remained consistently at that rate with a 95% confidence level of only +/- 0.09 mm/yr. (for perspective the average thickness of a human fingernail is about 0.42 mm) as shown in the NOAA data below. The longer the recorded period of tide gauge measurements at any location the smaller the 95% confidence levels.

Previous Relative Sea Level Trends
8518750 The Battery, New York

As more data are collected at water level stations, the linear relative sea level trends can be recalculated each year. The figure compares linear relative sea level trends and 95% confidence intervals calculated from the beginning of the station record to recent years. The values do not indicate the trend in each year, but the trend of the entire data period up to that year.

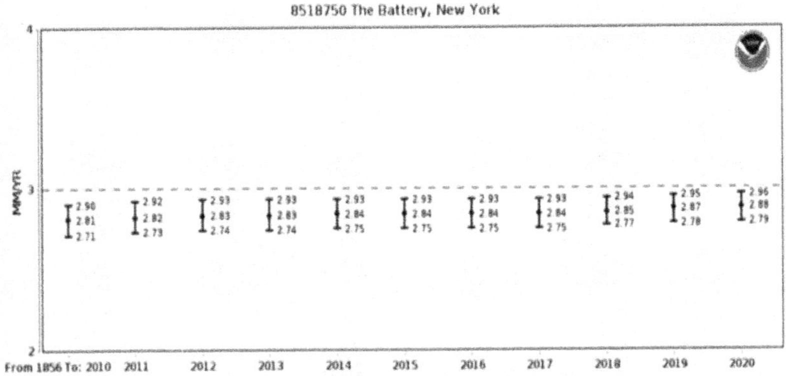

EXPORT TO TEXT | EXPORT TO CSV | SAVE IMAGE

Although the trend may vary with the end year, there is no statistically significant difference between the calculated trends if their 95% confidence intervals overlap. Therefore, the most recent calculated trend is not necessarily more accurate than the previous trends; it is merely a little more precise. If several recent years have anomalously high or low water levels, the values may actually move slightly away from the true long-term linear trend.

On the west coast of the U.S. the longest measured sea level rise tide gauge is located at San Francisco with a 123-year measurement record going back to the year 1897.

Relative Sea Level Trend
9414290 San Francisco, California

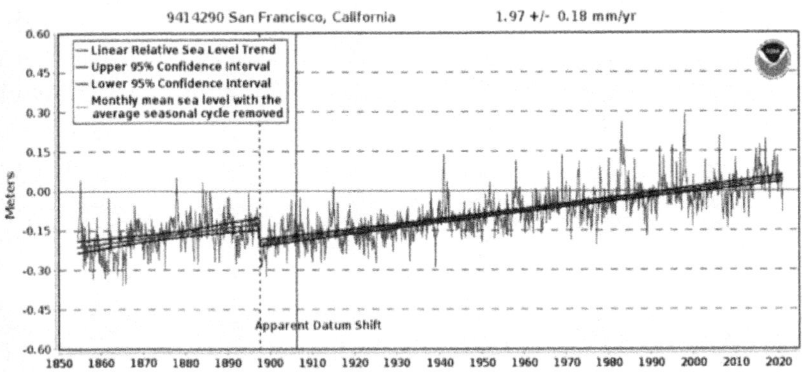

EXPORT TO TEXT | EXPORT TO CSV | SAVE IMAGE

The relative sea level trend is 1.97 millimeters/year with a 95% confidence interval of +/- 0.18 mm/yr based on monthly mean sea level data from 1897 to 2020 which is equivalent to a change of 0.65 feet in 100 years.

The rate of coastal sea level rise is 0.56 feet per century or 1.97 millimeters per year and has remained consistently at that rate with a 95% confidence level of only +/- 0.18 mm/yr. as shown in the NOAA data below.

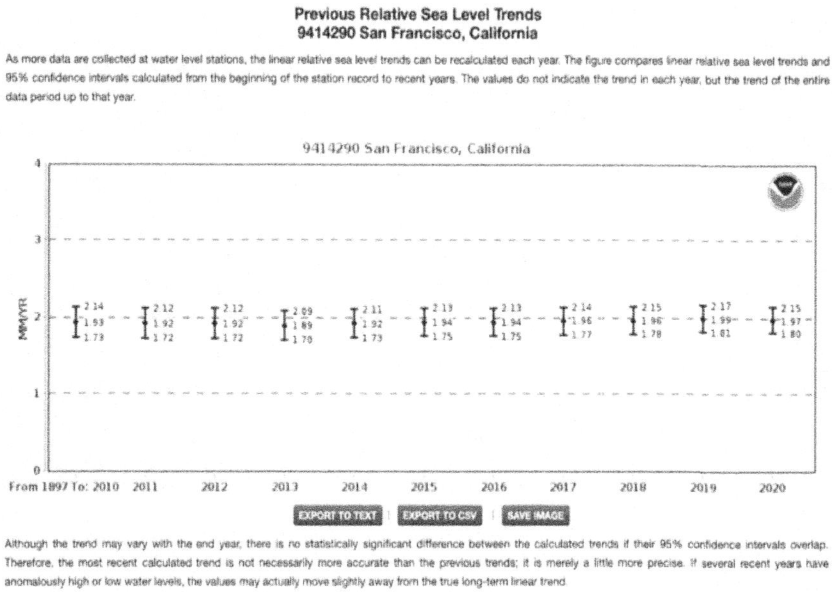

On the Hawaiian Islands the longest NOAA tide gauge data measurement station is located at Honolulu with a measurement record of 115 years going back to 1905.

Relative Sea Level Trend
1612340 Honolulu, Hawaii

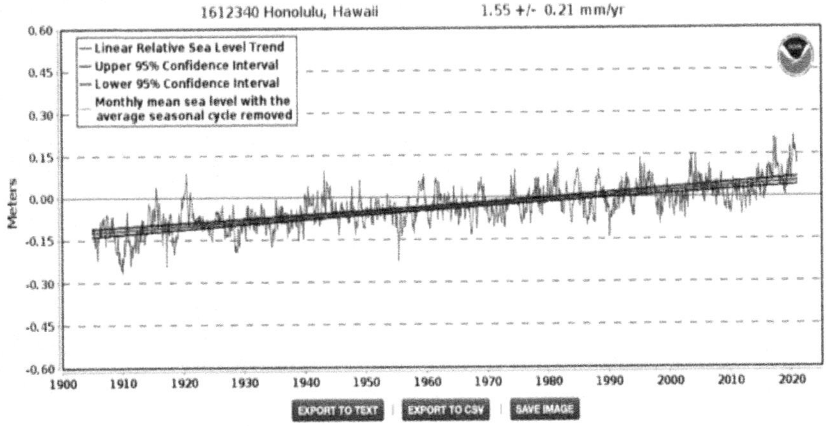

The relative sea level trend is 1.55 millimeters/year with a 95% confidence interval of +/- 0.21 mm/yr based on monthly mean sea level data from

The rate of coastal sea level rise is 0.51 feet per century or 1.55 millimeters per year and has remained consistently at that rate with a 95% confidence level of only +/- 0.21 mm/yr. as shown in the NOAA data below.

Previous Relative Sea Level Trends
1612340 Honolulu, Hawaii

As more data are collected at water level stations, the linear relative sea level trends can be recalculated each year. The figure compares linear relative sea level trends and 95% confidence intervals calculated from the beginning of the station record to recent years. The values do not indicate the trend in each year, but the trend of the entire data period up to that year.

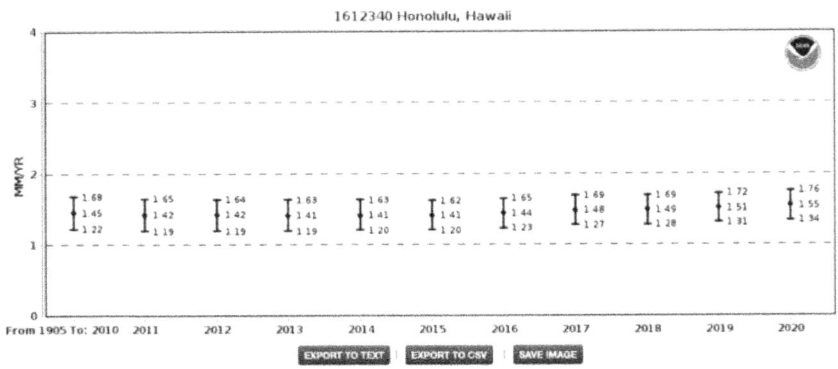

Although the trend may vary with the end year, there is no statistically significant difference between the calculated trends if their 95% confidence intervals overlap. Therefore, the most recent calculated trend is not necessarily more accurate than the previous trends; it is merely a little more precise. If several recent years have anomalously high or low water levels, the values may actually move slightly away from the true long-term linear trend.

Honolulu is the birthplace of President Obama who in June 2008 after securing the Democratic Party Presidential nomination made the claim that his election would be "the moment when the rise of the oceans began to slow and our planet began to heal."

Climate science informed individuals know the oceans have been rising gradually for thousands of years since the end of the last ice age and continued to do so after Obama's election. His misleading comments reflect the use of climate science deception and distortion as is the case for many Democratic Party politicians including those at the 1988 hearing.

NOAA tide gauge data updated through year 2020 measurements confirms that U.S. coastal sea level rise remain consistent and are not accelerating which is also the case for hundreds of other global tide gauge coastal locations under the worldwide GLOSS data measurement system.

Claims made at the Democratic Party's hearing in 1988 that the rate of sea level rise is climbing ever higher and threatens to rapidly inundate our coastline because of increasing greenhouse gas emissions have been proven WRONG.

The hearing then moved on to the statements and presentation of Dr. James Hansen and the NASA GISS climate model used to supposedly represent the behavior of the global average temperature anomaly based upon the level of greenhouse gas emissions.

Dr. Hansen's principal conclusions presented at the hearing are that the earth is warmer in 1988 than at any time in the history of instrumental measurements, that global warming is now sufficiently large that we can ascribe with a high degree of confidence a cause and effect relationship to the greenhouse effect and that the computer climate simulations show the greenhouse effect is already large enough to begin to affect the probability of extreme events such as summer heat waves with the model results implying that the heat wave/drought occurrences in the Southeast and Midwest of the U.S. may become more frequent in the next decade than climatological (1950-1980) statistics.

The EPA, NOAA and IPCC drought and heat wave data and analysis obtained over the past three decades as presented in a prior discussion clearly indicates that Dr. Hansen's principal conclusions regarding these events increased likelihood and severity from 1988 and beyond as a result of increasing greenhouse gas emissions are unsupported by actual data.

Dr. Hansen indicated that they have made initial studies with state-of-the-art climate models as noted in the diagram below where three different scenarios (as described previously regarding the failed future temperature increases noted by Senator Baucus of 1.5 to 4.5 degrees C by year 2030) of future increased levels of greenhouse gas emissions are used to attempt to portray the impact on the annual global average temperature anomaly of these different levels of emissions.

#54

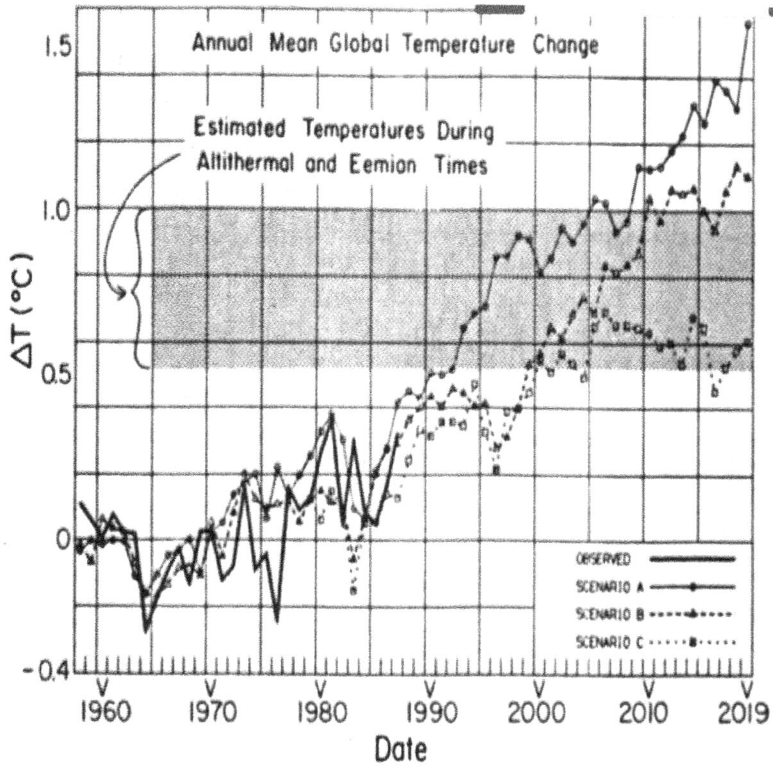

Fig. 3. Annual mean global surface air temperature computed for trace gas scenarios A, B and C described in reference 1. [Scenario A assumes continued growth rates of trace gas emissions typical of the past 20 years, i.e., about 1.5% yr^{-1} emission growth; scenario B has emission rates approximately fixed at current rates; scenario C drastically reduces trace gas emissions between 1990 and 2000.] Observed temperatures are from reference 6. The shaded range is an estimate of global temperature during the peak of the current and previous interglacial periods, about 6,000 and 120,000 years before present, respectively. The zero point for observations is the 1951-1980 mean (reference 6); the zero point for the model is the control run mean.

"Our studies during the past several years at the Goddard Institute for Space Studies have focused on the expected transient climate change during the next few decades."

The measured annual global average temperature anomaly data is now available for three decades after the 1988 hearings.

The graph below from a WUWT article shows the updated GISS (yellow) and HADCRUT (blue) measured global average temperature anomaly data since 1988 and also includes Dr. Hansen's original global average temperature anomaly data (red) from 1958 to 1986.

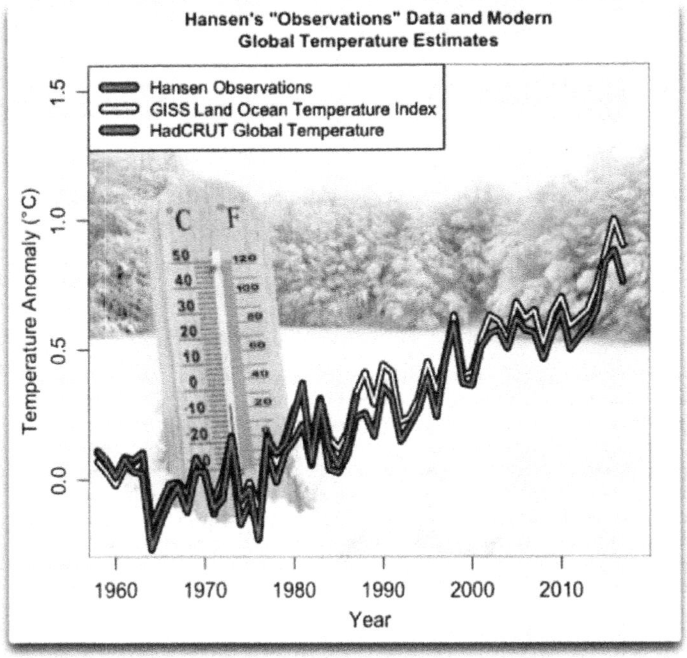

Figure 3. The line marked "Observations" in Hansen's graph shown as Figure 1 above, along with modern temperature estimates. All data is expressed as anomalies about the 1951-1980 mean temperature.

The next graph shows the measured global average temperature anomaly increases over the last three decades (yellow and blue) that clearly show that the NASA GISS computer model temperature anomaly estimates (purple) grossly overstates the impact of increasing greenhouse gas emissions on the global average temperature anomaly. Hansen Scenario A represents rapidly increasing greenhouse gas emissions that reflect the reality of continuing global emissions increases that have occurred since 1986 driven by huge emissions increases by the world's developing nations.

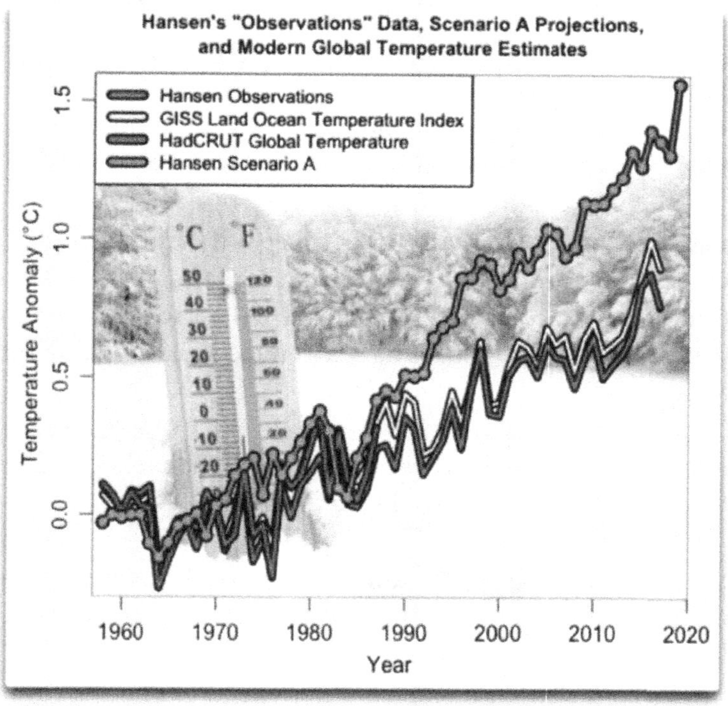

Figure 4. The line marked "Observations" in Hansen's graph shown as Figure 1 above, along with his Scenario A, and modern temperature estimates. All observational data is expressed as anomalies about the 1951-1980 mean temperature.

The significantly lower measured global average temperature anomaly data since 1986 shows a much closer relationship to Hansen's Scenario B and C lower temperatures projections which reflect significantly lower greenhouse gas emissions than have actually occurred.

There is nothing in these results that can confirm that man made actions are controlling global average temperature anomaly outcomes with these results just as likely reflecting natural climate variation driven outcomes.

It is clear that the 1988 hearing claims that increased greenhouse gas emissions will significantly increase future global average temperature anomaly outcomes have been grossly exaggerated and that the temperature increases presented in the hearing of 1.5 degrees C to 4.5 degrees C increase by year 2030 based on the NASA GISS computer model have been proven WRONG.

It is also clear that the real-world sensitivity of the global average temperature anomaly to increasing global greenhouse gas emissions has been greatly overstated with the computer model assumptions leading to this flawed conclusion also proven WRONG.

The next scientific "expert" presenter was Dr. Michael Oppenheimer, Senior Scientist, Environmental Defense Fund.

The Statements by Dr. Oppenheimer noted the following key points: "Global mean temperature will likely rise at about 0.6 degrees F per decade and sea level at about 2.5 inches per decade."

"These rates are about six times recent history."

"Furthermore, as long as greenhouse gases continue to grow in the atmosphere, there is no known natural limit to the warming short of catastrophic change."

"Because the oceans are slow to heat, there is a lag between emissions and full manifestation of corresponding warming, a lag which some estimate at 40 years."

"The world is now 1 degree F warmer than century ago and may become another 1 degree warmer even if conditions are curtailed today."

"If climate changes rapidly, agricultural and water resources will be stressed."

"Even if global food supplies are maintained, one need only look at the current Great Plains drought to see the human and economic cost associated with hot, dry weather in the grain belt. Weather of this sort we can expect with increasing frequency in the future." (The claim of increasing U.S. heat waves and droughts is unsupported by data as previously discussed.)

"Every decade of delay and implementation of greenhouse gas abatement policies ultimately adds perhaps a degree F of warming and no policy can be fully implemented immediately in any event."

"Slowing warming to an acceptable rate and ultimately stabilizing the atmosphere would require reductions in fossil emissions by 60% from present levels, along with similar reductions of other greenhouse gases."

Dr. Oppenheimer's claims are addressed as follows:
Prior information regarding the rate of increase of the global average temperature anomaly as determined by the UAH satellite measured data over the 33 years from 1988 to 2021 is 0.14 degrees C per decade or 0.252 degrees F per decade which is far below Dr. Oppenheimer's claim of 0.6 degrees F per decade increase.

NOAA tide gauge data updated through year 2020 (more than three decades after the 1988 Senate hearings) estimates that the global absolute rate of sea level rise is about 0.7 inches per decade (between 1.7 to 1.8 mm/yr.) which is strikingly below Dr. Oppenheimer's claim of 2.5 inches per decade increase.

Furthermore Dr. Oppenheimer's claim that each decade of delay in implementing greenhouse gas abatement policies adds another 1 Degree F of warming would mean that the three decades of continued growth in greenhouse gas emissions that have occurrence since the 1988 hearing would have considerably increased his estimates of 0.6 degrees F temperature anomaly increase per decade and 2.5 inches of sea level rise increase per decade.

The measured results (a three-decade long post hearing period) of the temperature anomaly increase (0.252 degrees F per decade) and sea level rise per decade (0.7 inches per decade) show the claim that large additional increased temperature anomaly and sea level rise outcomes will occur for each missed decade of greenhouse gas abatement delay are unsupported.

Global greenhouse gas emissions have INCREASED over the period of 1988 to 2019 by 67% (driven and controlled by the world's developing nations led by China and India) versus Dr. Oppenheimer's claim that a level of emissions abatement of 60% REDUCTION from 1988 levels is required to slow warming to an acceptable rate.

Measured data do not support Dr. Oppenheimer's claims of large increases of global average temperature anomaly and sea level rise occurring if greenhouse gases are not significantly reduced since the measured temperature anomaly and sea level rise increases were far below Dr. Oppenheimer's claims even **with increasing emissions.**

The claim that "there is no known natural limit to warming" as long as greenhouse gases are added to the atmosphere is incorrect. The greenhouse gas warming effectiveness diminishes logarithmically with increasing concentrations of greenhouse gases in the atmosphere.

Dr. Oppenheimer's claim that future global food shortages may occur because of increasing droughts and heat waves is unsupported by global food production data.

Global food production continues to climb and is at record levels. Global grain production in 2020-21 is forecast to increase to a record 2.224 billion metric tons according to the International Grain Council.

IGC projects record grain output in 2020-21

Photo: Adobe stock

03.29.2021 *By Arvin Donley*

LONDON, ENGLAND – Global grain production in 2020-21 is forecast to increase to a record 2.224 billion tonnes, according to the most recent grain market report from the International Grains Council.

Output in 2020-21 is expected to be 39 million tonnes larger than the previous year and 9 tonnes higher than last month's projection.

Record harvests of wheat and corn are also forecast at 774 million metric tons and 1.139 billion metric tons respectively. Soybean world production is also forecast up and just shy of the 2018-2019 record production level. Global rice production is also forecast to increase in 2020-21 to a peak of 504 million metric tons.

Dr. Oppenheimer's key climate claims presented at the 1988 hearing as discussed above are unsupported by subsequent data as follows:

Claims of increasing rates of global average temperature anomaly and sea level rise (0.6 degrees C per decade and 2.5 inches per decade sea level rise) due to increased greenhouse gas emissions are unsupported based on measured data covering the three-decade period from 1988 to 2021. These claims have been proven WRONG.

Claims that significant increases of rates of global average temperature anomaly and sea level rise will occur for each decade of delay in global greenhouse gas abatement have been proven WRONG.

The claim that there is no natural limit to future global average temperature anomaly increases due to greenhouse gas emissions has been proven WRONG.

Claims that global food shortages could occur because of global average temperature anomaly increases caused by increased greenhouse gas emissions have been proven WRONG.

The next presenter was Dr. George Woodwell, Director, Woods Hole Research Center.

Dr. Woodwell noted the following key points:

"We are embarked on a period of drastic climate change."

"We expect that that (doubling of carbon dioxide) would occur sometime early in the next century, 2030 or so."

"We expect the means (global average temperature anomaly) to run for the earth as a whole somewhere between 1 and a half and 5 and a half degrees C."

"The effect will be an increase in sea level of 30 cm to 1.5 m (about 1 foot to about 5 feet) over the next 50 – 100 years."

"We have the potential, as Dr. Oppenheimer just pointed out, of changing climate zones, altering the productivity of agricultural and changing the potential of earth for fixing green plants and changing it drastically."

"If warming proceeds rapidly enough to destroy forests (1 degree C per decade), that component can expand considerably."

Dr. Woodwell's claims are addressed as follows:
Previously discussed UAH satellite measured temperature anomaly data from 1988 to 2021 (33 years) shows a global average temperature anomaly rate of about 0.14 degrees C per decade or about 0.59 degrees C increase by year 2030 from 1988 versus Dr. Woodwell's claims of 1.5 to 5.5 degrees C increase from year 1988 to 2030.

Previously discussed NOAA absolute global sea level rise rate of 0.7 inches per decade means a 50-year increase to year 2038 of 3.5 inches and a 100-year increase to year 2088 of 7 inches versus Dr. Woodwell's claims of about 12 to 60 inches in the next 50 to 100 years. NOAA coastal tide gauge data updated to year 2020 at hundreds of locations are not showing acceleration of rates of coastal sea level rise.

Regarding potential reductions in global food production prior discussions document world record high food production levels of grains, wheat, corn, rice, soybeans, etc. Additionally, large population regions of the earth also reflect record high food production as noted in the items presented below addressing food production in China, India and South Africa.

CHINADAILY.COM.CN

Global Edition ASI
中文 双语 França

HOME CHINA WORLD BUSINESS LIFESTYLE CULTURE TRAVEL WATCHTHIS SPORTS OPINION RE

China News Society Innovation HK/Macao Cross-Straits Cover Story Photo

• Home / China / National affairs

Summer output of grains hits record high

By YANG WANLI | China Daily | Updated: 2020-07-16 08:55

A farmer uses a shovel to separate grains of wheat from the husk in Zhangyao village of Erlang town in Xiping county, Central China's Henan province on May 26, 2020. [Photo/Xinhua]

India's foodgrain production to be an all-time high at 303 million tonnes

Vishwa Mohan / TNN / Updated: Feb 25, 2021, 13:45 IST

FACEBOOK TWITTER LINKEDIN EMAIL AA

UP NEXT

1
India's foodgrain production to be an all-time high at 303...

2
More than one representation from a ward does not negat...

3
Vows for Eternity: A bespoke matchmaking servic...

4
India slams Organisation of Islamic Cooperation...

TRENDING TOPICS

- Corona cases in India
- Digvijaya Singh
- Maharashtra lockdown news
- Kumbh Mela 2021
- Delhi weekend lockdown
- Coronavirus India update live
- West Bengal Election
- NEET PG 2021 Postponed
- RR vs DC
- TS SSC Exams Cancelled

NEW DELHI: India's foodgrain production in the 2020-21 crop year will touch a new all-time record high of around 303 million tonnes (MT), which is over 2% higher than the previous year's output.

South Africa agriculture exports hit record on large harvest and efforts to keep sector open

News of the Day 29 November 2020 Hits: 288

- Wandile Sihlobo

These products will continue to support SA's agricultural exports in the final quarter of 2020.

Citrus has featured prominently and exports for 2020 are expected to reach a record 2.5-million tonnes, a 17% increase. Similarly, after the restriction of sales in domestic markets and disruption of exports during the lockdown period, wine exports could continue to improve in the last quarter.

This has all happened in a season in which SA's wine production was in recovery mode from a volume perspective, with the 2020 wine grape harvest estimated to be 8.2% higher than the previous year at 1.3-million tonnes. For maize too, the 2020/2021 exports are estimated at 2.5-million tonnes, a 35% annual rise due to the second-largest harvest on record. The same applies to other products that had already featured prominently in the exports list during the first three-quarters of the year due to large production volumes.

Earlier in the year, there was concern about logistical challenges at the ports and general uncertainty over global trade due to disruptions caused by the pandemic to supply chains and the debilitating effect of lockdowns on demand, but agricultural trade data show the sector has largely been insulated. On the domestic side, this is mainly due to joint efforts by the government and the private sector to ensure constant communication about challenges the industry faced, and action thereafter to resolve glitches.

Concerning negative impacts on global greening caused by increasing emissions of greenhouse gases NASA satellite observations reflect the opposite occurring as noted in the study below indicating that from a quarter to half of Earth's vegetation lands has shown significant greening over the last 35 years due to rising levels of CO_2.

POSITIVE EFFECTS OF CO2

NASA: Carbon dioxide fertilization greening Earth, study finds

5 years ago

Anthony Watts

From NASA/GODDARD SPACE FLIGHT CENTER – (we covered this in a previous release, but this press release brings new information – Anthony)

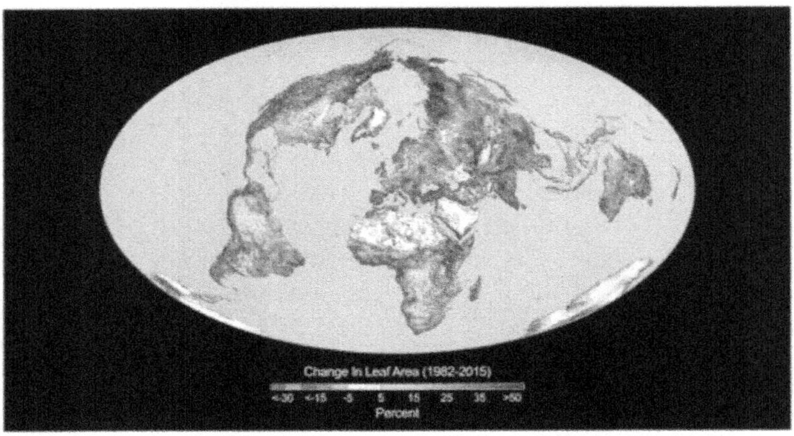

This image shows the change in leaf area across the globe from 1982-2015. CREDIT Credits: Boston University/R. Myneni

From a quarter to half of Earth's vegetated lands has shown significant greening over the last 35 years largely due to rising levels of atmospheric carbon dioxide, according to a new study published in the journal *Nature Climate Change* on April 25.

An international team of 32 authors from 24 institutions in eight countries led the effort, which involved using satellite data from NASA's Moderate Resolution Imaging Spectrometer and the National Oceanic and Atmospheric Administration's Advanced Very High Resolution Radiometer instruments to help determine the leaf area index, or amount of leaf cover, over the planet's vegetated regions. The greening represents an increase in leaves on plants and trees equivalent in area to two times the continental United States.

Regarding negative impacts of greenhouse gases on global forests a recent study by the University of Maryland shows otherwise as noted below.

"News headlines report a world constantly beset by deforestation and desertification, but new research suggests the planet may not be as tree-damaged as once thought.

Although agricultural expansion in the tropics has swallowed vast areas of the rainforest, climate change has allowed a greater number of new trees to grow in areas previously too cold to support them.

Scientists at the University of Maryland analyzed satellite pictures showing how the use of land on Planet Earth has altered over a 35-year period. The study, published in Nature journal, is the largest of its kind ever conducted.

The research suggests an area covering 2.24 million square kilometers – roughly the combined land surface of Texas and Alaska, two sizeable US states – has been added to global tree cover since 1982. This equates to 7% of the Earth's surface covered by new trees"

Additionally claims that greenhouse gas increases are driving more forest fires around the world have also been shown to be wrong based on NASA satellite data as noted below.

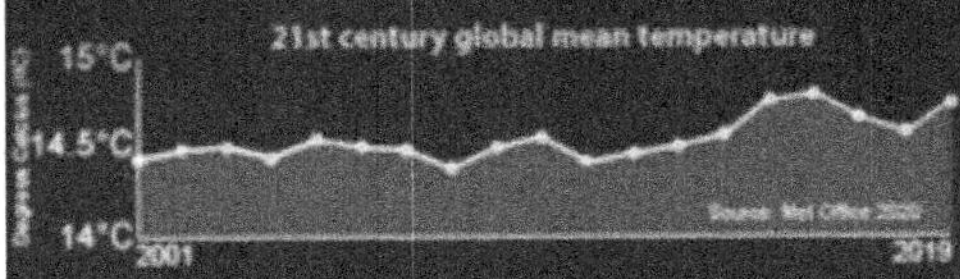

The Global Warming Policy Forum

Director: Dr Benny Peiser

Home | Who We Are | Latest Postings | Press Releases | Factsheets | GWPF Newsletter | Contact | RSS

You are here: The Global Warming Policy Forum (GWPF) » Satellite Observations Reveal Decreasing Trend in Global Wildfires

The Observatory

Satellite Observations Reveal Decreasing Trend in Global Wildfires

Date: 16/09/20 | GWPF

While wildfires in the Western U.S. continue to rage, satellite observations over the last 20 years have revealed a decreasing trend in global wildfires. What's going on?

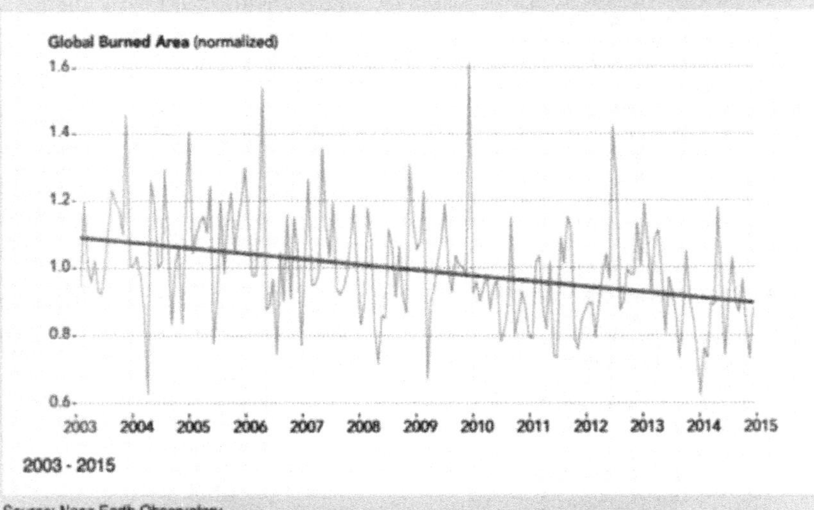

Source: Nasa Earth Observatory
https://earthobservatory.nasa.gov/images/145421/building-a-long-term-record-of-fire

As strong winds and hot air continue to propel wildfires across the Western U.S. states of California, Oregon and Washington state, politicians, activists and researchers quarrel violently about the main causes of these disasters and how to reduce the risk of wildfires in the future.

There can be little doubt that drought conditions and high temperatures are exacerbating these wildfires. However, over recent decades human activities such as land management and agriculture, increasing population density and active fire suppression have succeeded in significantly reducing the global areas burned by wildfires, despite the rise in global temperatures.

Regarding Dr. Woodwell's key comments we have the following conclusions:
Claims that the global average temperature anomaly will increase by 1.5 to 5.5 degrees C by year 2030 have been proven WRONG.

Claims that global sea level rise will further increase by 30 cm to 1.5 m in the next 50 to 100 years have been proven WRONG.

Claims that global agriculture production and world greening and forest growth will be negatively impacted by increasing greenhouse gas emissions have been proven WRONG.

There were a few other presenters at the hearing, but they basically continued to address the same type of claims as previously discussed.

Subsequent to the 1988 Senate hearing on global warming the UN IPCC has conducted a number of climate analysis reports and concluded that there are significant limitations to climate models being able to provide accurate future climate predictions that cannot be overcome.

The UN IPCC Third Assessment Report (AR3) was issued in year 2001 more than a decade after the 1988 hearing. That report finally acknowledged that it is not possible to develop climate models that can accurately model global climate and provide future climate predictions.

Specifically, the report in Section 14.2.2.2 noted:
"In sum, a strategy must recognize what is possible. In climate research and modeling, we should recognize that we are dealing with a coupled non-linear chaotic system, and therefore that the long-term prediction of future climate states is not possible. The most we can expect to achieve is the prediction of the probability distribution of the system's future possible states by generation of ensembles model solutions."

Claims at the 1988 Democratic global warming hearing that climate models can be developed to provide accurate assessments of future climate states have been proven WRONG.

UN IPCC Assessment Reports to date provide climate model scenarios (referred to as RCPs) that are used to suggest various possible climate states in the future. The RCPs that were included in the AR5 report issued in 2013 were identified as RCP 2.6, RCP 4.5, RCP 6.0 and RCP 8.5 with assumed greenhouse gas emissions varied from low scenario (2.6) to high scenario (8.5) along with numerous other assumptions.

However, these climate scenarios are all qualified as follows:
"The scenarios should be considered plausible and illustrative, and do not have probabilities attached to them." (12.3.1; Box1.1)

Climate models may serve useful purposes in academic and scientific studies but they are based upon conjecture and speculation since the RCP's climate scenarios utilized are simply "plausible" and "illustrative" and have no probabilities associated with their scenarios. These scenarios cannot be used to establish accurate assessments of future climate states.

The UN IPCC high carbon emission scenario RCP 8.5 has been challenged numerous times including by former chief scientist for Obama's Energy Department Steven Koonin for positing completely ridiculous and **implausible** assumptions of increases in coal use a century from now which is just one of dozens of assumptions thrown into these scenarios.

"A drumroll moment was Zeke Hausfather and Glen Peter's 2020 article in the journal Nature partly headlined: **"Stop using the worst-case scenario for climate warming as the most likely outcome."**

"This followed the 2017 paper by Justin Ritchie and Hadi Dowlatabadi asking why climate scenarios posit implausible increases in coal burning a century from now. And I could go on. Roger Pielke Jr. and colleagues show how the RCP 8.5 scenario was born to give modelers a high-emissions scenario to play with, and how it came to be embraced despite being at odds with every real-world indicator concerning the expected course of future emissions."

Yet this completely ridiculous scenario has been exploited by charlatan climate alarmist media such as the New York Times to manufacture unrealistic claims trying to mislead the country and its leaders that climate impacts in the future are much more severe than expected when in fact these claims are totally bogus and represent scientifically unsupported alarmist exaggerated propaganda.

"To this day, the print edition of the New York Times has never mentioned RCP 8.5, the unsupported emissions scenario on which so many of its climate jeremiads rest."

The UN IPCC has failed to develop climate analysis scenarios that have defined probabilities and instead simply invented scenarios using conjecture and speculation that have no scientifically defined event probabilities. These scenarios are simply tools for exploring guess work which climate alarmist schemers are misrepresenting to the public and exploited by Biden and Democratic Party politicians to push economically damaging and unnecessary climate legislation and mandates.

This is exactly what the Democratic Party 1988 Senate Hearings on Global Warming represented as exposed by the incredible number of flawed conclusions, claims and models as identified above.

In addition to the flawed climate scenarios the UN IPCC computer models are also flawed and have failed to represent global average temperature anomalies with accuracy as noted in the graph below which shows the UN IPCC models (referred to as CMIP5 generation) failing to project results anywhere close to measured global average temperature anomalies from 1979 through 2018.

On January 22, 2021, John Christy presented a new online talk to the Irish Climate Science Forum. The talk was arranged by Jim O'Brien. A summary of the presentation can be read at clintel.org here. In this post we present two interesting graphs from the presentation. These compare observations to the IPCC Coupled Model Intercomparison Project 5 climate models (CMIP5, 2013) and CMIP6 (current set of IPCC models) climate model projections.

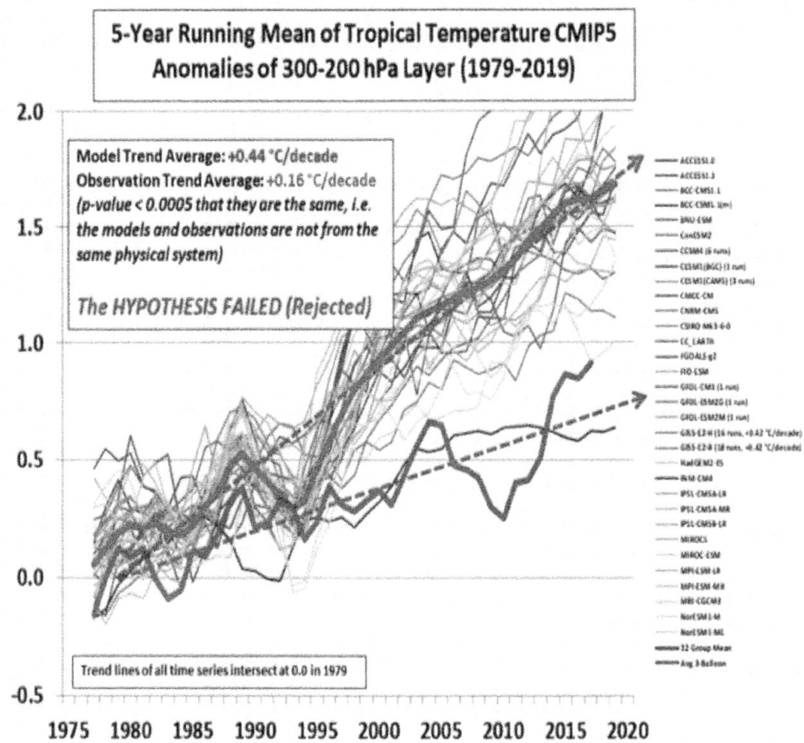

Figure 1. CMIP5 models versus weather balloon observations in green. The details of why the models fail statistically can be seen in a 2018 paper by McKitrick and Christy here.

The most recent climate models developed are identified by the UN IPCC as IPCC Coupled Model Intercomparison Project 6 models (CMIP6 generation) that utilize about 50 different sets of models (magically if you create 50 climate models where each model can't possibly be right and average them you end up with "robust" and useful outcomes according to climate alarmist scientists) to attempt to perform climate analysis forecasts of future climate outcomes.

A recent comparison of the effectiveness of the latest spread of UN IPCC model results trying to replicate the actual global average temperature anomaly measurements covering the period of 1979 to 2019 again shows very poor agreement with the actually measured global average temperatures anomaly data (dark green) during this period. Additionally, the CMIP6 models are increasingly divergent in temperature outcomes versus the CMIP5 models which is clearly moving in the wrong direction.

The differences between the UN IPCC model outcomes and measured temperature anomaly data are statistically significant and shows the models have been invalidated.

Use of UN IPCC climate scenarios and computer model outcomes are derived through use of conjecture and speculation and completely unsuited for purposes of regulatory driven mandates and commitments that require the expenditure of trillions of dollars of global capital which can be utilized for much greater benefit in dealing with known massive global problems including poverty, education and health care.

Climate alarmism started 33 years ago at the 1988 Democratic Party Senate hearings on global warming with these hearings clearly representing nothing but contrived speculation, conjecture, exaggeration, distortion, and deception in making scientifically unsupported claims in addressing climate issues.

The findings and conclusions of the 1988 Democratic Party Global Warming hearing have been proven WRONG.

The Democratic Party's pattern of climate alarmist distortion and deception has continued for more than three decades and is thriving under the incompetent energy leadership of Biden, the Democratic Party and their media shills who want to impose massive economic damage upon the country that will accomplish absolutely nothing regarding global climate impacts with natural climate variation much more likely controlling global climate outcomes.

ENDNOTES

CHAPTER 1

[1] United Nations Environment Programme (2012) The Emissions Gap Report 2012. A UNEP Synthesis Report, p. 2

[2] Overland, I., Sovacool, B. K. (2020). The misallocation of climate research funding. Energy Research & Social Science, 62.

[3] U.S. Government Accountability Office (2018). Climate Change: Analysis of Reported Federal Funding.

[4] Domonske, C., Dwyer, C. (2017). Trump Announces U.S. Withdrawal from Paris Climate Accord.

[5] Perry, M. J. (2019). Quotations of the Day from Friends of Science. Carpe Diem. AEI.org.

CHAPTER 2

[1] Editors. (n.d.). Quote by Albert Einstein, "No amount of experimentation can ever prove me right; a single experiment can prove me wrong." Brainy Quotes.

[2] Sargent, Jr., J. F., (n.d.). The U.S. Science and Engineering Workforce: Recent, Current, and Projected Employment, Wages, and Unemployment. Congressional Research Service. Federation of American Scientists.

3] Yoder, B.L. (n.d.). Engineering by the Numbers. American Association for Engineering Societies (ASEE).

[4] Editors. (n.d.). Bachelor's Degrees in Physics and STEM. American Physical Society (APS).

CHAPTER 3

[1] Cook, J. D., Nuccitelli, S. A., Green, M., et al. (2013). Quantifying the consensus on anthropogenic global warming in the scientific literature. Environ. Res. Lett. 8, 024024, 1-7.

[2] Mitchell, H., (2015) Yogi Berra quote: "no one goes there nowadays, it is too crowded." LA Times.

[3] Roberts, D. (2010). Behavior changes cause changes in beliefs, not vice versa. Grist.

[4] Schneider, S. (2011) The 'Double Ethical Blind' Pitfall. Stanford University: Mediarology.

[5] Perry, M. J. (2019). Quotations of the Day from Friends of Science. Carpe Diem. AEI.org.

[6] Twitter (2013) Barack Obama, "Ninety-seven percent of scientists agree: #climate change is real, man-made and dangerous."

[7] Editor. Global Warming: A Chilling Perspective. "William Gray Quote from Denver Rocky Mountain News."

[8] Editor. (2005). The Political Climate. Debating "Global Warming."

[9] Lindzen, R. S., (1997) Prepared Statement of Richard S. Lindzen, Alfred P. Sloan Professor of Meteorology, Massachusetts Institute of Technology. Committee on Environment and Public Works, United States Senate, 105th Congress, 1st Session, U.S. Government Printing Office pp113.

[10] Stanford University. (n.d.). Stephen Schneider, Ph.D. Biography. ClimateChange.Net.

[11] Memin. (n.d.). Tim Wirth (biography).

[12] The Montreal Millennium Summit. (n.d). Christine Stewart (Biography). Wayback Machine Internet Archive.

[13] Bell, L. (2013). In Their Own Words: Climate Alarmists Debunk Their 'Science'. Forbes.

[14] Curry, J.A. (2011). Stephen Schneider and the 'Double Ethical Bind' of Climate Change Communication.' Climate Etc.

[15] Rasool, S.I., Schneider, S.H. (1971). Atmospheric Carbon Dioxide and Aerosols: Effects of Large Increases on Global Climate. Science. 173, 3992, pp. 138-141.

[16] Massachusetts Institute of Technology (MIT). (n.d.). Earth, Atmosphere, and Planetary Sciences. Richard Lindzen (Biography). MIT EAPS Directory.

[17] Watts, A. (2017). Lindzen responds to the MIT letter objecting to his petition to Trump to withdraw from the UNFCC. Watts Up With That?

[18] Editor. (2017). The Einstein-Szilard Letter-1939. Atomic Heritage Foundation.

[19] Editor. (July). The Russell-Einstein peace manifesto-archive. The Guardian.

CHAPTER 4

[1] Helmholtz-Zentrum Berlin. (n.d.). The Helmholtz Climate Initiative.

[2] The Royal Society. (n.d.). Greenhouse gases affect Earth's energy balance and climate.

[3] University of Cambridge Centre for Climate Science. (n.d.). Research Themes.

[4] University of Oxford, Global Science Programme. (n.d.). Climate Projects.

[5] Institut De France. (n.d.). Académie des sciences. About Us.

[6] McNutt, M., Mote, Jr., C.D., Dzau, V.J. (2019). National Academies Presidents Affirm the Scientific Evidence of Climate Change. News Release. The National Academies of Sciences, Engineering, Medicine.

[7] Royal Society and U.S. National Academy of Sciences release joint publication on climate change. (2014). The Royal Society.

[8] NASA. (n.d.). Scientific Consensus: Earth's Climate is Warming. Global Climate Change: Vital Signs of the Planet.

[9] Richard, K. (2017). Russian Scientists Dismiss CO_2 Forcing, Predict Decades of Cooling, Connect Cosmic Ray Flux to Climate. NoTricksZone.

[10] Stozhkov, Y.I., Bazilevskaya, G.A., Makhmutov, V.S., et al. (2017). Cosmic Rays, Solar Activity, and Change in the Earth's Climate. Bulletin of the Russian Academy of Sciences: Physics, 80(2): pp 252-254.

[11] Katchev, G. (2019). Wildfires and Floods Push Russia to Revise Its Stance on Climate Change. Wall Street Journal.

[12] Editor. New Study Reveals the Atmospheric Footprint of the Global Warming Hiatus. Institute of Atmospheric Physics, Chinese Academy of Sciences.

[13] Liu, B., Zhou, T. (2017) Atmospheric footprint of the recent warming slowdown. Scientific Reports, 7, 40947.

[14] The Institute of Atmospheric Physics. Chinese Academy of Sciences.

CHAPTER 5

[1] IPCC. (n.d.). About the IPCC.

[2] Eschenbach, W. (2012). Defund the IPCC Now. Watts Up With That?

[3] IPCC (October 1990). FAR Climate Change: The IPCC Response Strategies.

[4] IPCC (April 2014). AR5 Synthesis Report: Climate.

[5] Liu, B., Zhou, T. (2017) Atmospheric footprint of the recent warming slowdown. Scientific Reports, 7, 40947.

[6] IPCC (2013). AR5 Report. Chapter 2SM – Observations: Atmosphere and Surface Supplementary Material [Online].

[7] NOAA. (n.d.). Climate Models. Climate.gov.

[8] IPCC (October 1995). SAR Climate Change 1995: Synthesis Report.

[9] IPCC (October 2001) TAR Climate Change 2001: Synthesis Report.

[10] IPCC (September 2007) Fourth Assessment Report (AR4).

[11] IPCC (October 2013). AR5 Report 2013: The Physical Science Basis: Chapter 9. Evaluation of Climate Models.

[12] IPCC (October 8, 2018). Special Report 15: Global Warming of 1.5°C.

[13] IPCC, 2019: Summary for Policymakers. In: Climate Change and Land: an IPCC special report on climate change, desertification, land degradation, sustainable land management, food security, and greenhouse gas fluxes in terrestrial ecosystems [P.R. Shukla, J. Skea, E. Calvo Buendia, V. Masson-Delmotte, H.- O. Pörtner, D. C. Roberts, P. Zhai, R. Slade, S. Connors, R. van Diemen, M. Ferrat, E. Haughey, S. Luz, S. Neogi, M. Pathak, J. Petzold, J. Portugal Pereira, P. Vyas, E. Huntley, K. Kissick, M. Belkacemi, J. Malley, (eds.)]. In press.

[14] Committee on Radiative Forcing Effects on Climate; Climate Research Committee; Board on Atmospheric Sciences and Climate Division on Earth and Life Studies (2005). Radiative Forcing of Climate Change: Expanding the Concept and Addressing Uncertainties. The National Academies of Sciences, Engineering, Medicine. National Academies Press, Washington, D.C.

[15] IPCC (May 1996). Revised 1996 IPCC Guidelines for National Greenhouse Gas Inventories.

[16] United Nations: Department of Economic and Social Affairs – Economic Analysis (September 3, 2016). UN/DESA Policy Brief #45: The nexus between climate change and inequalities. Department of Economics and Social Affairs.

[17] Perry, M. J. (2019). Quotations of the Day from Friends of Science. Carpe Diem. AEI.org.

[18] United Nations. (2015). Climate Change. The Paris Agreement.

[19] Pashley A. (2016). U.S. and China to sign Paris climate deal in April. The Guardian.

[20] Energy & Environment. (2017). Statement by President Trump on the Paris Climate Accord.

[21] Green Climate Fund.

[22] UN Treaty Database. (n.d.). Kyoto Protocol to the United Nations Framework Convention on Climate Change.

CHAPTER 6

[1] Editors. (n.d.) Scientific Method, defined. Brittanica.

[2] Thomson, W. Electrical Units of Measurement. In: Popular Lectures and Addresses. Cambridge Library Collection - Physical Sciences, Cambridge: Cambridge University Press. 2011; pp. 73-136. doi:10.1017/CBO9780511997242.006.

[3] Ayala, F. (2009). Darwin and the Scientific Method. Proceedings of the National Academy of Sciences of the United States of American. 106, 1, pp 10033-10039.

[4] Isaacson, W. Einstein-His Life and Universe. Simon and Schuster, U.K., 2007.

[5] Editors. (n.d.). Fair tests in physics: Examining eclipses. Understanding Science: How Science Really Works. Berkley.

CHAPTER 7

[1] Lucas, J. (2015]. What is Thermodynamics? Live Science.

[2] Dincer, I., Rosen, M.A. (2013). Thermodynamic Fundamentals. Chapter 1, Exergy (Second Edition). P. 1-20. London, Elsevier. Available: https://www.sciencedirect.com/science/article/pii/B978008097089 9000012.

[3] Lucas, J. (2016). What is the First Law of Thermodynamics? Live Science.

[4] Isaacson, W. Einstein-His Life and Universe. Simon and Schuster, U.K., 2007.

[5] Editors. (n.d.). Fair tests in physics: Examining eclipses. Understanding Science: How Science Really Works. Berkley.

6] Lucas, J. (2015). What is the Zeroth Law of Thermodynamics? Live Science.

[7] Milne, E. A. (2004). Yoshioka, Alan (Ed.). Fowler, Sir Ralph Howard (1889–1944), mathematical physicist and weapons researcher. London, Oxford University Press.

[8] Bellis, M. (2020). The History of the Refrigerator. ThoughtCo.

[9] Hmolpedia. (n.d.) Thermodynamics Quotes.

[10] Levenspiel O. (1984) The Three Mechanisms of Heat Transfer: Conduction, Convection, and Radiation. In: Engineering Flow and Heat Exchange. The Plenum Chemical Engineering Series. Springer, Boston, MA.

[11] Elert, G. Latent Heat. The Physics Hypertextbook.

[12] Elert, G. (n.d.). Sensible Heat. The Physics Hypertextbook.

[13] Blanchard, P., Devaney, R. L., Hall, G. R. (2006). Differential Equations. Thompson.

CHAPTER 8

[1] NASA. (n.d.). About GISS.

[2] Met Office Hadley Centre for Climate Science and Services. (n.d.). Climate Change.

[3] The National Space & Science Technology Center. University of Alabama Huntsville.

[4] Remote Sensing Systems.

[5] NASA Goddard Institute for Space Studies/GISTEMP Team. (2019). GISS Surface Temperature Analysis (GISTEMP).

[6] NCDC.NOAA. (n.d.). About the NCDC.NOAA.

[7] Hausfather, Z. (2014). Understanding adjustments to temperature data. Climate Etc.

[8] Skeptical Science (July 2015). What do the 'Climategate' hacked CRU emails tell us?

[9] IPCC (October 1990). FAR Climate Change: The IPCC Response Strategies,

[10] NCAR UCAR. (n.d.). Global Surface Temperature Data: HadCRUT4 and CRUTEM4. Climate Data Guide.

[11] IPCC (April 2014). AR5 Synthesis Report: Climate.

[12] NOAA. (n.d.). U. S. Historical Climatology Network. (USHCN)

[13] McLean, J. (October 2018). An Audit of the Creation and Content of the HadCRUT4 Temperature Dataset. Robert Boyle Publishing.

[14] Watts, A. (October 15, 2018). Met Office responds to HadCRUT global temperature audit. What's Up With That.

[15] Schiermeier, Q. (2009). Battle Lines Drawn Over Email Leak. Nature. 462, 551. doi:10.1038/462551an

[16] NASA. (July 20, 2017). Earth's Energy Budget

[17] Chambers, L. (NASA Langley Research Center) and Bethea, K. (SSAI). (2013). Energy Budget: Earth's most important and least appreciated planetary attribute. Astronomical Society of the Pacific, 84

[18] UCAR Center for Science Education (ECAR SciEd). About Us.

[19] Center for Science Education (ECAR SciEd). The Energy Budget.

[20] NASA. (October 21, 2014). Measuring the Earth's Albedo.

[21] NASA. (January 14, 2009]. Earth's Energy Budget. Earth Observatory.

[22] Lindsey, R., (January 14, 2009). Climate and Earth's Energy Budget. NASA. Earth's Observatory.

[23] Manley, R. and Reynolds, P. (n.d.). Climate Data Information. Albedo.

[24] NOAA (August 14, 2020). Climate Change: Global Temperature.

[25] Gregerson, E. Ed. (n.d.). Stefan-Boltzmann Law. Encyclopaedia Brittanica.

[26] Crepeau, J. (2008). Josef Stefan and His Contributions to Heat Transfer. 2008 Proceedings of the ASME Summer Heat Transfer Conference, HT 2008. 3. 10.1115/HT2008-56073.

[27] Clark, R. (2013). A Dynamic, Coupled Thermal Reservoir Approach to Atmospheric Energy Transfer Part I: Concepts. Energy & Environment, 24, (3/4), 319-340.

[28] Clark, R. (2011). The Dynamic Greenhouse Effect and the Climate Averaging Paradox. Ventura Photonics Monograph, VPM 001, Thousand Oaks, CA, Amazon

[29] Sharp, T. (2017). Earth's Atmosphere: Composition, Climate & Weather. Space.com.

CHAPTER 9

[1] Florida Atlantic University. (n.d.). Earth's Atmosphere. NASA: Climate Science Investigations.

[2] Hausfather, Z. (2008). The Water Vapor Feedback. Yale Climate Connections.

[3] Clark, R., (2019). The Greenhouse Effect. Ventura Photonics Monograph, VPM 001, Thousand Oaks, CA, Amazon

[4] Fourier, Joseph. (1822). Theorie Analytique de la Chaleur. Firmin Didot (reissued by Cambridge University Press, 2009; ISBN 978-1-108-00180-9)

[5] Chambers' Encyclopaedia. (1961). Arrhenius, Svante August. London, George Newnes, Vol 1, p 635.

[6] Svante Arrhenius. (1896b). On the Influence of Carbonic Acid in the Air upon the Temperature of the Ground, London, Edinburgh, and Dublin Philosophical Magazine and Journal of Science (fifth series), April 1896. vol 41, pages 237–275.

[7] Lockerby, P. (December 21, 2016). Carbon Cycles by Arvid Högbom. Science 2.0.

[8] IPCC (October 2013). AR5 Report 2013: Anthropogenic and Natural Radiative Forcing. Chapter 8. Evaluation of Climate Models.

[9] Clark, R. (2013). A Dynamic, Coupled Thermal Reservoir Approach to Atmospheric Energy Transfer Part I: Concepts. Energy & Environment, 24, (3/4), 319-340.

[10] Clark, R. (September 2019). Dynamic Climate Energy Transfer and the Second Law of Thermodynamics. Ventura Photonics Climate Note 7, VPCN 007.1, Ventura Photonics.

[11] Clark, R. (2010). A Null Hypothesis for CO_2. Energy & Environment, 21(4), 171-200.

[12] Coolman, R. (September 26, 2014). What is Quantum Mechanics? Livescience.

[13] Lexico. 'Spectroscopy' definition.

[14] Mehta, A. (2011). Introduction to the Electromagnetic Spectrum and Spectroscopy. Pharmaxchange.

[15] Editors. Quantum Quotes. Quantum Shorts.

[16] Editors. 'Quantum Mechanics' definition Lexico.

[17] Aaserud, F. (November 14, 2019). Niels Bohr. Encyclopaedia Britannica, Inc.

[18] Lindley, D. (2007). Uncertainty: Einstein, Heisenberg, Bohr, and the Struggle for the Soul of Science. New York, NY, Anchor Books

[19] Clark, R. (April 25, 2020). "What happens when CO_2 absorbs a photon?" Dr. Roy Clark, April 25, 2020, email to Guy K. Mitchell, Jr.

[20] Clark, R. (March 21, 2013). A Dynamic, Couple Thermal Reservoir Approach to Atmospheric Energy Transfer Part I: Concepts. Ventura Photonics.

[21] Clark, R. (2011). The Dynamic Greenhouse Effect and the Climate Averaging Paradox. Ventura Photonics Monograph, VPM 001, Thousand Oaks, CA, Amazon

[22] Editors. April 20, 2020). NOAA Weekly Mauna Loa CO_2. CO_2.Earth.

[23] Smirnov, B.M. (May 2, 2018). Collision and radiative processes in emission of atmospheric carbon dioxide. J. Phys. D: Appl. Phys. 51 214004

[24] UAH. (n.d.). Christy, J.R. University of Alabama in Huntsville. [Online].

[25] Spencer, R. (n.d.). About Dr. Roy Spencer. Roy Spencer, Ph.D., Climatologist, Author, Former NASA Scientist.

[26] The University of Alabama in Huntsville. (July 2020) Global Temperature Report. Maps and Graphs, UAH, The National Space Science & Technology Center, Earth System Science Center.

CHAPTER 10

[1] USGS. How Much Water is There on Earth?

[2] NOAA. How Deep is the Ocean? Ocean Facts. National Ocean Service.

[3] Jones, P.D., New, M., Parker, D. E., et al. (May 1999). Surface Air Temperature and Its Changes over the Past 150 years. Reviews of Geophysics, 37, 2, pp. 173-199.

[4] Minnett, P.J., Alvera-Azcárate, A., Chin, T.M., et al. (2019) Half a century of satellite remote sensing of sea-surface temperature. Remote Sensing of Environment. 233, 111366.

[5] Windows to the Universe. Temperature of Ocean Water.

[6] Dahlman, L. and Lindsey, R. Climate Change: Ocean Heat Content. NOAA.

[7] NOAA. Argo Float Program.

[8] Borunda, A. (August 14, 2019). Ocean warming, explained. National Geographic.

[9] IPCC UN IPCC, in the 5th Assessment Report published in 2013, entitled, "Observations: Ocean."

[10] NCDC NOAA. Global Climate Report 2014.

[11] Rasmussen, C. (July 9, 2015). NASA Finds Oceans Slowed Global Temperature Rise. NASA Jet Propulsion Laboratory. California Institute of Technology.

[12] NOAA. (August 10, 2017). International report confirms 2016 was warmest year of record for the globe.

[13] von Shuckmann, Karina & National Center for Atmospheric Research Staff (Eds.) (May 7, 2017). The Climate Data Guide: Ocean Heat content for 10-1500m depth based on Argo.

[14] Dahlman, L. and Lindsey, R. (February 13, 2020) Climate Change: Ocean Heat Content. Climate.gov.

[15] Argo Program.

[16] Roemmich, D., Alford, M.H., Claustre, H., et al. On the Future of Argo: A Global, Full-Depth, Multi-Disciplinary Array. Frontiers in Marine Science: Ocean Observation.

CHAPTER 11

[1] NOAA. (n.d.). What is the difference between weather and climate? NOAA. Historical Maps and Charts audio podcast. National Ocean Service.

[2] Turner, J.; Anderson, P.; Lachlan-Cope, T.; Colwell, S.; Phillips; Kirchgaessner, A. L.; Marshall, G. J.; King, J. C.; Bracegirdle, T.; Vaughan, D. G.; Lagun, V.; Orr, A. (2009). Record low surface air temperature at Vostok station, Antarctica (PDF). Journal of Geophysical Research. 114 (D24): D24102.

[3] Scambos, T. A.; Campbell, G. G.; Pope, A.; Haran, T.; Muto, A.; Lazzara, M.; Reijmer, C. H.; Van Den Broeke, M. R. (2018). Ultralow Surface Temperatures in East Antarctica from Satellite Thermal Infrared Mapping: The Coldest Places on Earth. Geophysical Research Letters. 45 (12): 6124 - 6133. Bibcode:2018GeoRL.45.6124S. doi:10.1029/2018GL078133. hdl:1874/367883.

[4] Post, E., et al. {December 4, 2019). The polar regions in a 2°C warmer world. Science Advances. Vol. 5, No. 12

[5] NASA, Global Climate Change. (August 9, 2020). Arctic Sea Ice Minimum. NSIDC/NASA.

[6] World Meteorological Organization. (June 23, 2020). Reported New Record Temperature of 38°C north of Arctic circle.

[7] Richner, H., and Hächler, P. (2013). Understanding and forecasting Alpine foehn. Mountain Weather Research and Forecasting: Recent Progress and Current Challenges, F.K. Chow, S.F.J. De Wekker, and B.J. Snyder, Eds., Springer, 219-260.

[8] Elviedge, A.D. and Renfrew, I.A. (2016) The Causes of Foehn Warming in the Lee of Mountains Bull. Amer. Meteor. Soc. 97 (3): 455–466.

[9] Canada's Polar Environments. Arctic Winds.

[10] Hansen, K. (2019). Warm winds trigger melting in Antarctica. April Warm Autumn Winds Could Strain Antarctica's Larsen C Ice Shelf. AGU Newsroom.

[11] Clark, R. (2020, August) The Atlantic Multi-Decadal Oscillation.

[12] Michon, S. (2020, April 28). Understanding climate: Antarctic Sea ice extent.

CHAPTER 12

[1] Cook, J. (May 16, 2013). Skeptical Science Study Finds 97% Consensus on Human-Caused Global Warming in the Peer-Reviewed Literature. Citizen Science Team. Skeptical Science.

[2] GAO. Climate Change Funding and Management. U.S. Government Accountability Office.

[3] Buchner, B., Trabacchi, C., Mazza, F., et al. (October 22, 2013). The Global Landscape of Climate Finance 2013. Climate Policy Initiative.

[4] Davis, J. H., Landler, M., Davenport, C. (September 8, 2016). Obama on Climate Change: 'The Trends are Terrifying.' New York Times.

[5] Editor. (December 20, 2018). Funding for Research Areas. National Cancer Institute.

[6] Helmholtz Association. Helmholtz Climate Initiative. Regional Climate Offices.

[7] Associated Press. (December 30, 2019). 'We must do everything humanly possible' to deal with climate change, Merkel urges.' Market Watch.

[8] Sheftalovitch, Z. (December 31, 2019). Merkel: Germany must do 'everything humanly possible' on global warming, Quote: 'It will be our children and grandchildren who have to live with the consequences of what we do or don't do today,' chancellor says." Politico.

[9] Mason, R. (December 11, 2017). Theresa May puts tackling climate change back on the Tory agenda. The Guardian.

[10] Lindzen, R. (November 30, 2019). Straight Talk About Climate Change. 30 Academic Questions. A Publication of the National Association of Scholars. ISSN 0895-48.

[11] Cole, S. and Buis, A. Editor: Karen Northon (Updated August 7, 2017). NASA Study Finds Indian, Pacific Oceans Temporarily Hide Global Warming.

[12] Casey, J.L. Medieval Warming Period. The Grand Solar Minimum.

[13] Mann, M.E., Zhang, Z., Rutherford, S., et al. (2009). Global Signatures and Dynamical Origins of the Little Ice Age and Medieval Climate Anomaly (PDF). Science, 326 (5957):1256-60. Doi:10.1126/science.1177303.

[14] Payne, R. E. (March 6,1972). Albedo of the Sea Surface. Woods Hole Oceanographic Institute, Journal of the Atmospheric Sciences, 29.

[15] Editor. The Importance of Understanding Clouds. NASA. NASA Facts.

[16] Argo Program.

[17] Editor. (January 17, 2003). The Inconstant Sun. NASA.

[18] Australian Space Academy. The Solar Constant.

[19] USGS. (2020). The Sun and Climate. USGS Science for a Changing World, USGS Fact Sheet FS-095-00, August 2000.

[20] Macdougall, D. Milutin Milankovitch. Serbian mathematician and geophysicist. Britannica.

[21] Editor. (2019). The polar regions as components of the global climates system. World Ocean Review.

[22] Editor. (March 24, 2000). Milutin Milankovitch (1879-1958). NASA Earth Observatory.

[23] Hays, J.D., Imbrie, J., Shackleton, N.J. (December 10, 1976). Variations in the Earth's Orbit: Pacemaker of the Ice Ages. Science. 194 (4270), pp. 1121-1132.

[24] NASA. (March 24, 2000). For about 50 year's Milankovitch's theory was largely ignored. NASA Earth Observatory.

[25] National Archives. Federal Registers, (2021, February 1). Tackling the Climate Crisis at Home and Abroach. https://www.federalregister.gov/documents/2021/02/01/2021-02177/tackling-the-climate-crisis-at-home-and-abroad

ABOUT THE AUTHOR

GUY K. MITCHELL, JR. was born in a small town in Alabama in 1948. He was educated in the public school systems in Birmingham, AL, and attended the University of Alabama-Tuscaloosa, where he graduated with a Bachelor's of Science Degree in Mechanical Engineering. He worked in various positions in his career, including as a shift foreman in a coal-fired power plant, a plant superintendent in a steel mill, a general manager of a construction materials company in Saudi Arabia and Senior-Vice President of a Fortune 500 company.

Mr. Mitchell founded Mitchell Industries Inc. in 1996 and is the Chairman of DeSHAZO, a wholly owned subsidiary of Mitchell Industries. DeSHAZO is a leading manufacturer of heavy-duty industrial overhead cranes and automated equipment employing robotics; in addition, the Company provides aftermarket parts and services for overhead cranes and robotic systems throughout the U.S. and Mexico.

Mr. Mitchell was elected a member of Pi Tau Sigma International Mechanical Engineering Honor Society in December 1969 and graduated with a B.S.M.E. from the University of Alabama in May 1972. In May 1989, he completed the Advanced Management Program at Harvard Business School. In April 1995, he received the "Department of Mechanical Engineering Distinguished Fellow Award" from the Faculty of the Department of Mechanical Engineering and was elected a "Distinguished Engineering Fellow" by the College of Engineering, University of Alabama, the same year.

ABOUT THE AUTHOR

Mr. Mitchell became interested in the man-made global warming hypothesis in 2017. He has spent the last four years conducting extensive investigations into scientific research dealing with the subject, as well as the reports of the United Nations Intergovernmental Panel on Climate Change. In addition, he has spent several years studying spectroscopy, quantum mechanics and atmospheric physics to understand how the first principles of science in each of these fields impacts the analysis of the global warming hypothesis.

He and his wife live in Vero Beach, FL and have three children and six grandchildren. He enjoys hunting, fishing, playing golf, collecting wine, and playing the drums.

Made in United States
North Haven, CT
24 May 2022

19477943R00163